# Phylogenetic Analysis
and Paleontology

# Phylogenetic Analysis and Paleontology

Proceedings of a symposium entitled
"Phylogenetic Models,"
convened at the North American Paleontological
Convention II,
Lawrence, Kansas, August 8, 1977

edited by
Joel Cracraft and Niles Eldredge

Columbia University Press
New York

Library of Congress Cataloging in Publication Data

Main entry under title:

Phylogenetic analysis and paleontology.

   Includes bibliographies and index.
   1. Phylogeny—Congresses.  2. Paleontology—
Congresses.  3. Evolution—Congresses.  I. Cracraft,
Joel.  II. Eldredge, Niles.
QE721.P47     575     78-31404
ISBN 0-231-04692-8
ISBN 0-231-04693-6 pbk.

Columbia University Press
New York and Guildford, Surrey

Copyright © 1979 Columbia University Press
All rights reserved

Printed in the United States of America
6    5    4    3    2

# Contents

Introduction to the Symposium     1
    N. Eldredge and J. Cracraft

Phylogenetic Analysis, Evolutionary Models, and Paleontology     7
    J. Cracraft

Stratophenetic Approach to Phylogeny Reconstruction in Vertebrate Paleontology     41
    P. D. Gingerich

An Introduction to the Logic of Phylogeny Reconstruction     79
    E. S. Gaffney

Recognition of Ancestor–Descendant Relationships in Invertebrate Paleontology     113
    S. S. Bretsky

Cladism and Common Sense     165
    N. Eldredge

Cladistics: Is It Really Different from Classical Taxonomy?     199
    A. J. Boucot

Ancestors, Species, and Cladograms—Remarks on the Symposium     211
    E. O. Wiley

Index     227

# Symposium Participants

Arthur J. Boucot — Department of Geology, Oregon State University, Corvallis, Oregon 97331

Sara S. Bretsky — Department of Earth and Space Sciences, State University of New York, Stony Brook, New York 11794

Joel Cracraft — Department of Anatomy, University of Illinois at the Medical Center, P.O. Box 6998, Chicago, Illinois 60680

Niles Eldredge — Department of Invertebrates, The American Museum of Natural History, Central Park West at 79th Street, New York, New York 10024

Eugene S. Gaffney — Department of Vertebrate Paleontology, The American Museum of Natural History, Central Park West at 79th Street, New York, New York 10024

Philip D. Gingerich — Museum of Paleontology, The University of Michigan, Ann Arbor, Michigan 48109

E. O. Wiley — Division of Fishes, Museum of Natural History, The University of Kansas, Lawrence, Kansas 66045

# Phylogenetic Analysis
# and Paleontology

# Introduction to the Symposium

**Niles Eldredge and Joel Cracraft**

There has been an unprecedented amount of introspection among systematists in recent years. As especially reflected in the pages of *Systematic Zoology*, systematists have been examining the theoretical basis of their science. The pursuit has been relentless and, at times, merciless as old shibboleths are put aside and new ones proposed in their stead.

It would be natural to assume that paleontologists have been in the thick of things as the debate on how to reconstruct life's history rages on. Nothing could be further from the truth. Paradoxically, the working days of most paleontologists, whatever they might be filled with, are not devoted to drawing cladograms, phylogenetic trees, or any other similar diagrams depicting their views of phylogenetic relationships among taxa. The prime goal of paleontology is not the reconstruction of life's history—at least judging from the activities of the vast majority of its practioners. Contrary to the remarks of a reviewer of an early version of the present book, the entire multivolumed work *Treatise on Invertebrate Paleontology* contains relatively little about phylogenetic relationships—in many cases absolutely nothing (e.g., the volume of trilobites)—and this can be taken as a fairly accurate reflection of the way paleontologists typically order their priorities.

Despite this state of affairs, the myth persists among paleontologists (and also quite a few of our neontological colleagues) that fossils somehow afford the best, if not the only, means of reconstructing phylogenetic history. This view has been challenged persistently since Darwin's day. Haeckel, for example, never believed it. It stands to reason that if paleontology affords a special, privileged means of insight into evolutionary history, its practitioners have an obligation to specify their method. But, with few exceptions, they have been collectively silent during most of the recent discussion about method and theory in phylogenetic reconstruction.

The above perception of paleontological practice provided the rationale for the present symposium, which was held as part of the North American Paleontological Convention II, convened at the University of Kansas in Lawrence, Kansas, on August 7–10, 1977. All participants were chosen for their conviction that phylogenetic analysis is important in paleontology and that attempts to systematize fossil species should be based on an analysis of phylogenetic history.[1] All appear also to believe that good systematics underlies all but the most purely theoretical work done in paleontology. Hence the reconstruction of phylogeny assumes a position of central importance as the basis for subsequent research on fossils—after all, how can we do anything with a fossil unless we know what it is? The participants in this symposium appear to agree that "What it is" in the last analysis boils down to "What are its relationships?" It is our collective intention in this symposium to underscore our conviction that phylogenetic research has been badly neglected in paleontology and that such research of the highest caliber is necessary for further progress in our understanding of the nature of life's history. We test hypotheses of process with corroborated hypotheses of pattern, but we do not yet have more than a rather dim and confused notion of that pattern.

A second, more specific aim of this symposium is to expose and elucidate the contemporary controversy regarding the methodology of phylogenetic reconstruction. All of the participants have a strong paleontological back-

---

[1] Numerical taxonomy is conspicuous in its absence in this symposium—all the more surprising, perhaps, because the symposium was held in Lawrence, Kansas, long the North American center of numerical taxonomy. The decision to omit numerical taxonomy from the symposium reflects the disinterest in phylogenetic reconstruction expressly stated by the leading theoreticians of numerical taxonomy. However, numerical techniques are of course germane to phylogenetic reconstruction, and more importantly, a considerable segment of systematists believe that phenetics—the ordering of the biota in terms of overall phenotypic similarity—is important for the estimation of overall genetic similarity and hence crucial to phylogenetic reconstruction. Thus phenetics, the theoretical basis of numerical taxonomy, is represented in several contributions to this symposium.

## Introduction 3

ground and were invited on the basis of their recent contributions to the debate on these and related topics within paleontology. The plan of the book follows the organization of the symposium; there are five papers (which comprised the formal session) followed by two shorter papers commenting on the formal ones. The five formal papers were organized as a sort of debate. The first paper, by J. Cracraft, presents an overview of phylogenetic research in paleontology—its history, relationship to evolutionary theory, current status, and proper place within the discipline. The remaining four papers are organized to illustrate the diversity of opinion within vertebrate paleontology (P. D. Gingerich and E. S. Gaffney) and invertebrate paleontology (S. S. Bretsky and N. Eldredge). Each of the latter four speakers was given a list of questions (devised by Cracraft and Eldredge, organizers of the symposium) which he or she was asked to consider during preparation of the manuscript: What, exactly, is a phylogeny? What is the relationship between evolutionary theory, on the one hand, and ideas about the nature of phylogenetic reconstruction, on the other? What are ancestors and how might they be recognized? What is character analysis? Is there any point in recognizing a formal distinction between primitive and derived character states? And if so, how do we determine what is primitive and what is derived? In connection with character analysis, how do we utilize intrinsic data (i.e., comparative anatomical and morphogenetic information) and how do we handle extrinsic data (i.e., information on the distribution of organisms in space and time)? and finally, how do we string all of this information together into a phylogeny?

These questions were formulated, not to serve as an outline for each paper, but to give common points of reference on the more crucial issues currently debated, thus providing a common theme to each of the latter four papers. "Standard," or "traditional," paleontological approaches to phylogenetic reconstruction have been only occasionally, and in general incompletely, developed in the literature. Opinions differ, of course, as to exactly how this approach should be structured. In this symposium, P. D. Gingerich and S. S. Bretsky develop their own original formulations which, although they transcend "traditional" approaches in many respects, nonetheless appear to represent, in analytic form, prevailing and preferred views on the methodology of phylogenetic reconstruction in paleontology. E. S. Gaffney and N. Eldredge, on the other hand, discuss their views on the application of "phylogenetic systematics" (also known as "cladism") to the fossil record. No two cladists agree with each other (or, for that matter, with Willi Hennig) on every point, and this "school" of systematics is no more a monolith than that of the more traditional "evolutionary taxonomy," or even the school of "numerical taxonomy." Thus the papers were organized as a sort of debate, with the essential difference being that there were no formal presentations explicitly criticizing the opposing point of view, in the expectation that the two invited critical commentaries and the open discussion would supply ample quantities of criticism.

The five formal papers were presented at the opening session of the North American Paleontological Convention II on August 8, 1977. The two invited commentaries, by A. J. Boucot and E. O. Wiley, were presented in the evening of August 8 and were followed by several hours of discussion. All seven manuscripts were prepared before the convention and circulated to the other participants; the commentaries by Boucot and Wiley were prepared in direct response to the drafts of the five papers presented formally. Final versions of all seven manuscripts were prepared in response to commentary from other symposium participants, persons in attendance at the symposium, and reviewers arranged for privately by the individual authors and Columbia University Press. On the whole, they are faithful renditions of the gist, if not the exact verbiage, of the symposium held in Lawrence.

Was the symposium a success? Judging from the attendance at both the formal session in the morning and the more informal evening session, it was. Our basic aim was to stimulate interest in the general subject of phylogenetic reconstruction in paleontology and to inform anyone interested as to what the issues, in fact, are. All of the participants received a great deal of commentary on the material they presented, and in some cases, the commentary was not very polite or flattering. But commentary it was—hence our conclusion that the desired stimulation was achieved.

The title of the symposium, "Phylogenetic Models," turned out to be, in a sense, a *double entendre*. Many people interpret "phylogenetic models" as sets of views pertaining to the nature of the evolutionary process. Nearly all of the invited participants discussed evolutionary theory, and many of the evening discussants seemed frankly more interested in such notions than in the nitty-gritty of phylogenetic reconstruction. We acknowledge the stronger appeal that evolutionary theory (i.e., the causal process itself) may have over the seemingly drier subject of methodology of phylogenetic reconstruction (i.e., analysis of historical pattern). But surely these subject areas are related. Some of us think that notions of the evolutionary process logically imply which of the available approaches to phylogenetic reconstruction should be chosen. Some of us think that we can intelligently discuss evolutionary processes within paleontology only when we compare existing hypotheses about process with phylogenetic patterns and thus that the pattern must be worked out before statements on process can be made. Whatever the proper connection may be, evolutionary theory cannot progress much further without a better set of patterns of historical relationships among the elements of the biota, and it is to this subject that the book is specifically addressed. Perhaps by the time of North American Paleontological Convention III we shall be ready for a symposium on the relationship between paleontology and evolutionary theory per se. For now, we are pleased to present the proceedings from the symposium on phylogenetic reconstruction held in Lawrence in 1977.

# Acknowledgments

We wish to thank Dr. Roger Kaesler, of the University of Kansas, who chaired the organizing committee of the Convention and aided us greatly in our organizational and procedural tasks. We thank the invited participants for their cooperation in supplying information, outlines, and various versions of their manuscripts. Finally, we also thank those who participated in the informal evening discussion, thereby adding further dimensions and points of view to those already expressed by the seven invited participants.

# Phylogenetic Analysis, Evolutionary Models, and Paleontology

## Joel Cracraft

The last 10 to 15 years have produced major advances in the theory and methodology of the historical sciences—geology and biology. These advances have increased our understanding of the history of the earth and the history of life upon the earth. In geology, a new conceptual framework, plate tectonics, has compelled us to interpret old knowledge in new ways and to seek new knowledge via questions generated by this revolutionary perspective—questions impossible to ask within the framework of older theories. Similarly, within biology, new vigor has been infused into systematics—that discipline of biology most concerned with the history of life—again because of the challenge of new theory and methodology. Clearly, the innovations in geology have been profound; they have been equally so in biology, extending to all branches of systematics, including phylogenetic analysis, speciation theory, classification, and biogeography. Through the years, much has been made of the results of the "New Systematics" during the 1940s and the early 1950s; it is my view that we are experiencing an even more significant period in the study of historical biology. Nowhere, perhaps, is that change going to be felt more than in paleontology.

In this paper, I want first to trace certain aspects of the history of phylogenetic analysis within biology and paleontology. Many of the ideas about phylogenetic methodology prevalent within paleontology during the latter half of the twentieth century are derivable from philosophical changes that took place in the seventeenth and eighteenth centuries and were carried over into the nineteenth century, when Darwin gave evolutionary theory its first major exposition. Phylogenetic analysis is used here in a broad sense, to include both the reconstruction of historical pattern (phylogeny) and the study of the causes of that history (evolutionary processes). Some contemporary paleontologists may not see a necessary connection between phylogenetic reconstruction and a specific causal mechanism; it is certainly possible to reconstruct a pattern without knowledge of the underlying causes of that pattern. Some paleontologists believe this, and for the most part, so do I. But this is not the point I wish to address. Rather, it is that many paleontologists have adopted a research strategy that is, at least subconsciously, based squarely on their preconceptions about the nature of species and how they evolve. Furthermore, to ignore this connection between causal mechanism and phylogenetic research strategy constrains one's view to that of historical hindsight. Historically, discussions about the actuality of evolution, and how it might have occurred, preceded attempts to formalize a method to reconstruct pattern. At the center of those early discussions was the debate over the discreteness and reality of species, a debate that continues to influence strongly paleontological thinking and methodology. An examination of how these historical trends in thought have come to pervade much present-day paleontology and neontology is instructive in identifying the strengths and weaknesses of modern thinking.

In the final sections of this paper, I want to suggest that the adoption of the theory and methodology of phylogenetic systematics and the allopatric speciation model offers paleontology an important conceptual framework for analyzing the history of life. These two approaches are at the heart of the current debate about paleontological theory, and their adoption by paleontology may be imminent if one takes as evidence what can be observed now in the writings of theoretically oriented paleontologists and neontological systematists. No one could doubt that the concepts of phylogenetic systematics and allopatric speciation have already made a strong impact. At the very least, they have provided the impetus for reexamining the efficacy of more traditional approaches.

# Historical Perspective of Phylogenetic Analysis

The origins of current paleontological thinking about phylogenetic analysis must be considered against the background of the development of modern science. The philosophical confrontations extending from the Greeks to the present continue to influence our personal epistemological viewpoint, that is, how we believe understanding (knowledge) of nature is obtained. So it was with the eighteenth- and nineteenth-century biologists, including Darwin. The thesis developed here is not unique, perhaps, but a summary statement of the origins of phylogenetic thinking within paleontology has not, to my knowledge, been made. My assertion here is twofold: First, Darwin's perception of the evolutionary process was founded not only on the scientific evidence he compiled but also on what seems to have been his intellectual capture by certain philosophical precepts that developed after the Renaissance (indeed, his evidence and these precepts were often in conflict). Second, post-Darwinian paleontology, extending up to the present, merely assimilated Darwin's view, with minor modifications (this second point was also made by Eldredge and Gould, 1972). Neither of these assertions should be startling; much of science has traditionally operated this way.

## SOME PRE-DARWINIAN INFLUENCES

The seventeenth and eighteenth centuries were a time of profound alteration in mankind's conceptions of the physical universe and the geological and biological worlds. Several important beliefs had been inherited from the Greek and medieval writers. These beliefs were taken for granted and formed the template for interpretations of natural phenomena. From the Platonic "principle of plenitude"—that the diversity of kinds of objects in the universe can be exhaustively exemplified—and the Aristotelian notion that things are continuous, the acceptance of the Great Chain of Being came to dominate much intellectual activity (Lovejoy, 1936). The objects of nature, including living organisms, were thus conceived as a chain, arranged in a hierarchical order of perfection, in which the kinds of objects differed "from that immediately above and that immediately below it by the 'least possible' degree of difference" (Lovejoy, 1936, p. 59).

The Aristotelian belief that natural motion existed toward the center of the universe—the earth—but elsewhere was possible only under the influence of a constant moving cause implied a general stability to the universe and its constituents; "The Aristotelian doctrine of inertia was a doctrine of rest" (Butterfield, 1957, p. 15). Furthermore, the Unmoved Mover of Aristotle eventually opened the door for ecclesiastical didactics—after all, something, somewhere, must keep things moving.

During the sixteenth and seventeenth centuries, the works of Copernicus, Galileo, Newton, and others contributed to the replacement of the traditional Aristotelian idea of a more or less stable, fixed universe with a view of constant change and motion governed by a few fundamental natural laws. Although the mechanical materialism of Newton pointed to the concept of change in nature, the spectre of Scripture remained a burden, particularly to those concerned not so much with the celestial manifestations of God's wisdom as with more earthly themes. A fundamental truth permeating all thought of the time was the act of Creation—supernaturalism had not yet been supplanted entirely by naturalism, even in the minds of mechanistic astronomers—and creationism itself implied a "stable framework of structures" (Greene, 1961, p. 23).

With the heavens seemingly operating under the influence of natural causes, as manifested in the natural laws of the physical sciences, and God being relegated more and more to the position of First Cause, the time had come to search for natural causes here on earth. By the latter half of the eighteenth century, scientists were ascribing earth history to natural phenomena and, in the process, two fundamental preconceptions were being threatened. The first challenge was derived from empirical observation of nature in action. Change of the earth's surface appeared to be slow, and unless modification could be ascribed to catastrophic forces, one was compelled to envision vast amounts of time to accomplish that change. The second challenge, which is coupled with the first, occurred when the idea of catastrophism—so readily understandable to those searching the rocks for confirmation of the Deluge—was gradually replaced by the notion of uniformitarianism, which could not be accepted without an increased time scale.

Both the concept of uniformitarianism and the significant observations of Cuvier and William Smith that fossil forms are found in superimposed strata not of the same age contributed to the perception that organisms, as well as rocks, had had a long history. That different strata contained different organisms (few of which were still extant) suggested the possibility of change, but more importantly, change that was very gradual.[1] In assessing the intellectual milieu at this time, Lovejoy (1959, p. 364) remarks: "Enough time had now been provided for extremely gradual changes in the characters of successive generations of the descendants of any pair of organisms; and any plausible hypothesis of the natural descent of, e.g., extant species from ancestors from which they differed widely in form or functions, necessarily presupposed such gradualness."

The conviction of gradual change arose from two sources within eighteenth- and early nineteenth-century science and philosophy. The first, just noted, was the logical correspondence between the slow gradual change of the earth's geological formations and the presumed slow change of the organisms housed therein. The second source came about with the replacement of the philosophical doctrine of essentialism by a nominalistic view of the *scala naturae*.

The philosophy of essentialism was derived from the writings of Plato and Aristotle and is characterized by the belief that knowledge is only advanced once the true nature of things has been discovered and described. "In Aristotle's view three things can be known about any entity—its essence, its definition, and its name. . . . The Name names the essence. . . . The definition gives a complete and exhaustive description of the essence" (Hull, 1965, p. 6). Entities that had an essence or Form were called species, and

> for Plato and Aristotle and for centuries of philosophers to follow, species names referred primarily to these metaphysical entities and not simply to classes of things. It was these metaphysical entities which were *really* real. One of the most important transitions in Western philosophy was the shift from such things as essences and Forms being the basic constituents of the world to the objects of experience being the basic constituents. This shift was reflected in the word "species" in biology (Hull, 1967, pp. 310–11).

---

[1] I am not arguing here that uniformitarianism necessarily implies evolution. Although Lovejoy (1959, p. 366) seems to accept this position, others point out that uniformitarianism was antievolutionary because it advocated cyclical change (Mayr, 1972, pp. 984–85; Gale, 1972, p. 334). Uniformitarianism did, however, imply slow gradual change.

The dominant view of species, then, particularly after the rise of Christian theology, was that they were immutable, fixed entities in nature. That species might change or become extinct was contrary to the belief in the perfection of nature; those species produced at the time of the Creation were expected to exist until God's final dissolution. The fixity and stability of God's world, and therefore of species, was not a question of interest, it was a revealed truth. This world view found its highest expression in the classification attempts of Linnaeus, who sought an orderly arrangement of the natural relationships of organisms, these relationships being believed to be the direct manifestation of the divine pattern of nature itself. But before long this divinely inspired order became, if not chaos, at least a good deal more complicated than that envisioned by Linnaeus.

The breakdown of essentialism grew out of a long philosophical reaction to Aristotelianism, and perhaps had its greatest impetus in the philosophy of Leibniz and Locke. To them, essences were surely definable by God and possibly by angels but not by mere human beings (Lovejoy, 1936, pp. 228–29).

Within natural history, as systematic collections grew and new plants and animals were discovered, classification became increasingly difficult. There developed the perception among naturalists that groups once easily separable were linked by intermediate forms. Distinctions disappeared; the pattern of nature was seen to be one of fine gradations from species to species. The most well-known naturalist to adopt this view was Buffon. In his early writings, he challenged the Linnaean species as being only an artifact because only individuals, not species, were real. Nature was a chain of insensible gradations among forms. In his later writings, Buffon dropped his nominalistic conception of species and adopted one close to the present-day biological species concept (Hull, 1967; Sloan, 1976). The notion of fine gradations, however, persisted.[2]

---

[2] Although most present-day biologists probably view the introduction of biological evolution as a distinctly mid-nineteenth-century event, significant erosion of the belief in the fixity of species was taking place nearly a century earlier. By the 1740s, Linnaeus himself was suggesting that some plant species were transmutations of previously existing species (Burkhardt, 1977, p. 77; see also Toulmin and Goodfield, 1977, p. 177). That there was considerable discussion about the origins of new species is evidenced by a comment of the French botanist Michel Adanson in 1769: "One of the most celebrated and discussed questions in the past several years in natural history and especially in botany is to know if the species among the plants are constant, or if they change; that is, if by sexual reproduction or otherwise, new species are formed which become set in their own right, reproducing themselves constantly under this new form without reverting to that of the plants from which they originated" (Adanson, 1769, *Mem. Acad. Roy. Soc.;* cited from Burkhardt, 1977, p. 77).

For Lamarck, fine gradations among organisms had an evolutionary meaning. Except for gaps not yet filled, Lamarck believed that "neighbouring species and even genera are separated by the finest differences" (Lamarck, 1914, p. 38). Furthermore,

> as the individuals of one of our species change their abode, climate, habits or manner of life, they become subject to influences which little by little alter the consistency and proportions of their parts, their shape, properties and even their organisation . . . thus, after a long succession of generations these individuals, originally belonging to one species, become at length transformed into a new species distinct from the first (1914, pp. 38–39).

Although the Aristotelian concept of distinct and stable species retained advocates, most notably Cuvier and Lyell (Coleman, 1964), the notion of imperceptible differences among species gained acceptance. Whereas geological observations made change gradual in a temporal sense, the destruction of essentialism and the rise of nominalism facilitated acceptance of the view that change in form from one species to another must also be viewed as gradual. Thus this idea, coupled with those of slow change and vast amounts of time, became part of the intellectual viewpoint that Darwin eventually adopted and extended.[3]

## DARWIN'S PALEONTOLOGICAL PERSPECTIVE

Charles Darwin, although a scientific revolutionary by most standards, was captured by prior scientific and philosophical tradition and operated comfortably within it. Perhaps he could not have done otherwise; it is rare for a novel scientific idea to gain acceptance unless presented within the framework of contemporary thought, and total radicalism is just as unsuccessful, if not more so, in science as in other areas of human endeavor.

As is well known, Darwin was an uncompromising gradualist. He saw the

---

[3] Toulmin and Goodfield (1977, p. 141) note the following: "During the second half of the eighteenth century, philosophical debate had shown the possibility of replacing the static Renaissance view of Nature by a dynamic, developmental one. But philosophy alone could not dispatch the older time-scale finally: for that, an alternative numerical chronology had to be worked out from the evidence of scientific observation." This view reinforces the ideas developed here, i.e., that consideration of the origins of Darwin's evolutionary thinking before 1859 requires attention to the philosophical, as well as scientific, milieu existing at the time.

origin of species as a slow process: "we have every reason to believe the process of manufacturing new species to be a slow one" (1859, p. 56; all quotes from first edition unless specified otherwise); "The process of modification and the production of a number of allied forms *must* be slow and gradual" (1859, p. 317; italics added). And he saw evolution as occurring over long periods of time: "A man must for years examine for himself great piles of superimposed strata, and watch the sea at work grinding down old rocks and making fresh sediment, before he can hope to comprehend anything of the lapse of time. . . ." (1859, p. 282).

What is less appreciated about Darwin's paleontological perspective is the reason he had that view in the first place. Darwin, it is often said, was successful in advancing the concept of evolution because he marshaled vast amounts of supporting data. Relative to his contemporaries, that may be true, but one searches in vain through his two chapters on the geological record for evidence to support his contention that evolution was a slow, gradual process, one of insensibly fine gradations between ancestors and descendants. Indeed, the "factual information" that Darwin presents (there was virtually none) seems to support a philosophical (and scientific) viewpoint opposite to that of his own. Darwin was the consummate theorist, a scientist of the highest stature who did not let data stand in the way of his ideas.[4] If this interpretation of Darwin is correct, it may have something important to say about the origins of the theory and methodology of nineteenth- and twentieth-century paleontology.

Darwin ascribed the absence of "innumerable transitional links" to the "extreme imperfection of the geological record." As an aside, one is led to ask to what degree the "imperfect" geological record was viewed as such by Darwin simply because there were no observed transitional links. There can be no doubt that Darwin had empirical reason to believe in an imperfect record, but was his emphasis based partly on his biological expectations? In any case, there is no question that Darwin *expected* those finely graded transitions. He firmly believed that transitions must have existed and struggled to rationalize their absence:

[4] Lovejoy (1959) has argued that the "facts" supporting evolution were known many years prior to 1859. Hence it might be concluded that "facts" had little to do with the acceptance of evolution and that evolution was not, at least initially, a factual question at all: "Darwin's major role was not to produce an adequate theory of evolution, indeed he failed to do this, but rather to jettison altogether the appeal to the Bible and to rely entirely upon experience and non-teleological, non-Providential explanations" (Ruse, 1970, p. 16).

So must the number of intermediate varieties, which have formerly existed on the earth, be truly enormous (1859, p. 280).

We should always look for forms intermediate between each species and a common but unknown progenitor (1859, p. 280).

It cannot be doubted that the geological record, viewed as a whole, is extremely imperfect; but if we confine our attention to any one formation, it becomes more difficult to understand, why we do not therein find closely graduated varieties between the allied species which lived at its commencement and at its close (1859, pp. 292–93).

On the theory of descent, the full meaning of the fact of fossil remains from closely connective formations, though ranked as distinct species, being closely related, is obvious. As the accumulation of each formation has often been interrupted, and as long blank intervals have intervened between successive formations, we ought not to expect to find . . . in any one or two formations all the intermediate varieties between the species which appeared at the commencement and close of these periods; but we ought to find after intervals, very long as measured by years, but only moderately long as measured geologically, closely allied forms, or, as they have been called by some authors, representative species; and these we assuredly do find. We find, in short, such evidence of the slow and scarcely sensible mutation of specific forms, *as we have a just right to expect to find* (1859, p. 336; italics added).

Darwin's expectations and theorizations were clearly at variance with his observations of nature. This was particularly evident for two related concepts. The first is his expectation of graduated transitions just noted. He expected transitions but could not find them in the rocks, saying: "Geology assuredly does not reveal any such finely-graduated organic chain; and this, perhaps, is the most obvious and gravest objection which can be argued against my theory. The explanation lies, as I believe, in the extreme imperfection of the geological record" (1859, p. 280). Darwin was also faced with the problem of not finding a transition within any single formation (see the third quotation in the preceding paragraph, but this, he reasoned, could be understood if the length of time it took to deposit a formation was shorter than the time it took for a species to evolve—again, species evolution being viewed as very slow.

Second, although he did not believe in discrete species, his observations of nature indicated the opposite—that they did exist. Darwin argued against discrete species in a theoretical context:

> Supposing B and C to be two species, and a third, A, to be found in an underlying bed; even if A were strictly intermediate between B and C, *it would simply be ranked as a third and distinct species, unless at the same time it could be most closely connected with either one or both forms by intermediate varieties* (p. 297; italics added).

> Unless we obtained numerous transitional gradations [between a parent-species and its descendants], we should not recognize their relationships, and *should consequently be compelled to rank them all as distinct species* (p. 297; italics added).

Of course, Darwin saw the evidence for species distinctness as a major problem:

> One [objection to the theory], namely the distinctness of specific forms, and their not being blended together by innumerable transitional links, is a very obvious difficulty (p. 297).

> Geological research . . . has done scarcely anything in breaking down the distinction between species, by connecting them together by numerous, fine, intermediate varieties; and this not having been effected, is probably the gravest and most obvious of all the many objections which may be urged against my views (p. 299).

An incomplete geological record became useful, and necessary, to explain away the paleontological observations which Darwin took to be at variance with his theory. Included among these were the sudden appearances of species in the fossil record, not to be expected with slow gradual transition but acceptable assuming the critical fossiliferous formations had been destroyed by the vicissitudes of time.

The fossil record, seemingly so important for anyone advocating evolutionary modification through time, was not very kind to Darwin's cause. As a result, he ignored it; the fossil record certainly did not make him alter his theorizations or expectations. In fact, what he often saw was stasis. Recall his already cited remarks about not finding change within any single formation. He further stated, in a passage added to later editions of *The Origin of Species,* that "the period during which each species underwent modification, though long as measured by years, was probably short in comparison with that during which it remained without undergoing any change" (1872, p. 320).

In summary, Darwin's paleontological perspective was that of a slow, gradual, *in situ* change of one species into another. [See Vorzimmer (1970, pp. 159–85) for a discussion of Darwin's views on isolation and migration in the origin of species.] A complete explanation of the origins of this perspective may not be possible, but one cannot help being impressed with the similarities between Darwin's position and the results of changes that occurred within philosophy and science during the seventeenth and eighteenth centuries.

The conflicts between Aristotelian essentialism and the ideas of philosophers such as Leibniz and Locke had profound effects on biological thought (Lovejoy, 1936). Were species discrete or were they imperceptibly merged with one another? To believe the former carried the connotation of creationism, that is, considering a species to be "a whole independent of number, independent of time; a whole always living, always the same; a whole which was counted as one among the works of the creation, and therefore constitutes a single unit in the creation" (Buffon, *Histoire Naturelle,* 1749; cited in Lovejoy, 1936, p. 230). This widely held philosophical stance was apparently appreciated by Darwin: "No one definition has yet satisfied all naturalists; yet every naturalist knows vaguely what he means when he speaks of a species. Generally the term includes the unknown element of a distinct act of creation" (1859, p. 44). Perhaps it is unreasonable to expect a person who labored for years on a scientific idea that would nail the coffin shut on creationism to be willing to adopt a philosophical belief that itself was a fundamental supposition of creationism. Clearly, Darwin saw species as mutable, and because he could not accept major saltative changes ("Natural selection can act only by taking advantage of slight successive variations; she can never take a leap, but must advance by the shortest and slowest steps"; 1859, p. 194), he was left only with gradualism. And gradualism, coupled with the mechanism of natural selection, was slow, orderly, and progressive—a philosophically more comfortable position could hardly be imagined for a respectable Victorian scientist.[5]

To summarize, the fundamental assertions of Darwin's paleontological perspective are as follows: (1) evolution is an exceedingly slow process; (2) the transformation of one species into another is gradual not only temporally but also morphologically; and (3) species are not discrete but arbitrary seg-

---

[5] This interpretation should not be taken as a criticism of Darwin's position. Since most biologists of the eighteenth and early-nineteenth centuries equated discrete species with creationism and immutability, Darwin's line of argumentation is reasonable. Later, with the development of speciation theory, it became possible to modify this view to permit the acceptance of discrete species and evolutionary change.

ments of an evolutionary continuum. Because it seems that the *Origin* contained very little empirical support for these assertions, Darwin's axiomatic acceptance of them probably was in large part what might be called "extrascientific" [see Gale's comments (1972) on the "struggle for existence"]. Whereas the belief that evolution was slow could be inferred as a corollary of geological processes being slow, Darwin's ideas about gradual change and the discreteness ("reality") of species were embedded in the general philosophical thinking of the times. Coleman (1962, p. 326), in paraphrasing Lyell's dilemma with evolution, states the problem as it must have been seen by Darwin and his contemporaries: "If the species are not real, the obvious consequences are alarming: unlimited change becomes not only possible but even necessary. Species will no longer have well-defined limits, classification becomes a purely arbitrary exercise, and any species may easily be transformed into another." Thus it would seem that the philosophical beliefs which drove many contemporary biologists away from evolution and toward acceptance of discrete species also helped determine the world view of Darwin and later nineteenth-century evolutionists.

## THE POST-DARWINIAN PALEONTOLOGICAL PERSPECTIVE

The history of paleontological thought since the publication of *On The Origin of Species* is distinctly Darwinian in tone. The slow, gradual change of one species into another has been accepted as the predominant mode of species origination. Eldredge and Gould (1972) and Gould and Eldredge (1977) have provided examples of this gradualistic conception within paleontology, and even a cursory examination of the literature demonstrates its pervasive influence.

The rise of descriptive paleontology in the last half of the nineteenth century perhaps had as its major goal (aside from the mere description of organic diversity) the verification of the Darwinian view of life's history. Zittel, one of the more influential paleontologists, wrote that "It often happens that species and genera occurring in a given formation reappear in the next following with scarcely any perceptible changes, so that the doctrine of the gradual transformation and transmutation of older forms is irresistibly forced upon one" (1913, p. 12). Furthermore, he continues, "According to the theory of descent no sharp specific distinctions can be drawn" (1913, p. 13). [Eldredge and Gould (1972, p. 91) cite additional passages from Zittel showing a gradualistic viewpoint.]

At the turn of the century, few paleontologists dominated the field as did Henry Fairfield Osborn at The American Museum of Natural History. Osborn was straightforward in his advocacy of documenting morphological change through time rather than confirming species transformations: "The *origin of new characters*, not the origin of species, is the central problem in evolution" (Osborn, 1910). This outlook, although neglecting the question of the origin of species, engendered a paleontological methodology that continues to this day. Osborn was dead set against saltation and maintained an undeniable position for gradualism:

> The paleontologist can never demonstrate a discontinuity, because he must always entertain the suspicion that it may arise from lack of evidence.
>
> The paleontologist can never prove that the appearance of a new type combining a *number* of adaptive characters is a sudden appearance, because he must always admit the possibility that such a type may have slowly evolved elsewhere and come into the field of his observation suddenly through migration (Osborn, 1910).

Osborn regarded the "hypothesis of gradual appearance," as he called it, to be "the greatest contribution which paleontology has made to evolution." There can be little doubt that succeeding generations of paleontologists have felt scant urge to challenge this conceptual framework or its *modus operandi*.

The growth of evolutionary biology in the first half of the twentieth century was the primary result of advances in evolutionary genetics and species-level systematics. Whether the hoped-for synthesis of the "new systematics" was achieved is a matter for debate, but a number of books laid the foundation for a new approach to evolutionary problems (e.g., Dobzhansky, 1937; Huxley, 1940, 1942; Mayr, 1942). As a general statement, it is questionable whether this important body of theory and methodology had much effect on the thinking, or the practice, of paleontology. The most notable exception, of course, was the work of George Gaylord Simpson (1944, 1953), who made a monumental attempt to incorporate genetics and neontological systematics into paleontology.

A detailed analysis of Simpson's efforts is not necessary here. It will suffice to say that he attempted to synthesize viewpoints that often had premises fundamentally opposed to one another. Not unexpectedly, Simpson's main allegiances fell on the side of paleontological tradition, and he remained essentially a Darwinian gradualist. Simpson's analyses are extremely complex, and my desire is not to reduce them to a few summary statements if that means a misrepresentation of his position. Rather, I want to document, in Simpson's own writings, a position that has had an extraordinary influence on contemporary paleontology, particularly the study of fossil vertebrates. Moreover, Simpson's position will exemplify a number of issues to be discussed later in this paper.

Simpson's view of speciation is basically a modified case of phyletic gradualism in which splitting has occurred: "Speciation is the process of differentiation *within* populations and of the rise of genetic isolation between populations formally part of the same species. *Subsequent divergence is not, strictly speaking, speciation but an aspect of phyletic evolution*" (1953, p. 380; italics added). Simpson characterizes speciation either as the process of obtaining genetic isolation between two populations or as the "origin of new characters and their distribution or differentiation among and within populations" or both (1953, p. 379). Most neontologists would not define speciation in terms of morphological differentiation (e.g., Mayr, 1963), but they might include differentiation as a possible result of speciation or use morphology to recognize species-level taxa. Although he recognizes geographic isolation of populations as a contributing factor to speciation, Simpson clearly does not emphasize its importance to the extent that neontologists do. Basically, Simpson sees speciation in terms of the slow divergence of subsets of a continuous population, much as Darwin himself did.

Simpson (1953, p. 384) distinguishes between diversification and progression in evolution and links these to the two evolutionary modes, speciation and phyletic evolution: "When an ancestral species splits into two or more descendant species by segregation of existing variation, phyletic evolution is not an essential part of the picture; and when a single population undergoes extensive, cumulative change, there is no splitting."[6]

---

[6] My interpretation of Simpson's view is as follows: The process of isolation does not involve significant morphological change—that only comes later with phyletic evolution within isolates. It is not immediately apparent to me why there can be no splitting in a population undergoing extensive change. Simpson's position can be contrasted with that of Eldredge and Gould (1972),

Simpson's perception of phyletic evolution (gradualism) is standard:

> If a single lineage is being considered and if it changes so that the descendant and ancestral populations differ about as much as is usual between two related contemporaneous species, we say that one species has given rise to another. This, the essence of change in phyletic evolution, is at least as fundamental for evolution in general as is the separation of two species from what was originally one (1953, p. 386).

> Phyletic evolution is particularly evident to paleontologists and the most abundant examples from the fossil record illustrate this aspect of evolution more clearly than any other (1953, p. 385).

> An evolutionary species is defined as a separate lineage . . . of unitary role. If you start at any point in the sequence and follow the line backward through time, there is no place where the definition ceases to apply (1961, p. 165).

> If the fossil record were complete, you could start with man and run back to a protist still in the species *Homo sapiens*. Such classification is manifestly both useless and somehow wrong in principle. Certainly the lineage must be chopped into segments for purposes of classification, and this must be done arbitrarily . . . because there is no nonarbitrary way to subdivide a continuous line (1961, p. 165).

In summary, Simpson's approach to the origin of species is distinctly paleontological and does not correspond especially closely with the neontological conception. His concept of phyletic evolution is traditional, although cloaked in more modern language than the ideas of his predecessors or most of his contemporaries. Finally, his notion of species in evolutionary time is that they are arbitrary segments of a continuum, and this is certainly an orthodox paleontological viewpoint (see, for example, Sylvester-Bradley, 1956).

It is unnecessary to add further documentation of the modern paleontological perspective of the evolutionary process; this has been done adequately by Gould and Eldredge (1977). What remains is to discuss the efficacy of this perspective, how it accords with other evidence about the evolutionary process, and what its implications are for reconstructing the history of life.

---

because in the former, morphological change is the result of slow, directional selection, whereas in the latter, change is concentrated around the time of isolation and is not the result of long periods of directional selection. This dichotomy of viewpoints remains one of the more interesting questions in paleontology and is in need of empirical investigation.

# Paleontology and the Origin of Species

## THE ARGUMENT: WHICH MODEL DO WE ADOPT?

The current argument in paleontology about the origin of species contrasts the classical idea of phyletic gradualism (phyletic evolution) and the punctuated-equilibrium (allopatric speciation) model of Eldredge (1971, 1972) and Eldredge and Gould (1972). There have been criticisms (particularly Gingerich, 1976; Hecht, 1974) and responses (Eldredge and Gould, 1974; Gould and Eldredge, 1977), and most of the discussion has focused not so much on theory as on the validity of possible examples of each mode. It seems to me that Gould and Eldredge (1977) have marshaled a good case for claiming that virtually all of the *very best* examples of gradualism lack firm empirical confirmation. Others in this symposium will challenge this statement, and I will not enter into the debate on that level. Rather, I want to take a somewhat different approach to the study of the origin of species, particularly within paleontology, an approach that will amplify and expand the conclusions of Eldredge and Gould.

My discussion centers around the following question: Which mode (model) of the evolutionary process is most likely to lead to generalizations within evolutionary biology, to unification of evolutionary paleontology and neontology, and to advancement of paleontology itself? Thus, although we must and should be concerned with the soundness of empirical support for each model, the choice of a particular model or theory at any point in time should also be dictated by factors outside the realm of detailed empirical confirmation. This is not a radical proposal by any means, since most scientific theories, especially initially, are accepted or rejected on this basis, and data play a far less important role than might first be assumed. Just as Darwinism was accepted by the biological community for reasons other than empirical support, nonempirical considerations are pertinent in the controversy over the paleontological view of evolutionary modes.

I would identify three main "nonempirical" reasons for preferring the adoption of the allopatric speciation model in paleontology over that of phyletic gradualism (phyletic evolution):

1. **The allopatric speciation model unifies paleontology and neontology in terms of the conceptual framework predominating within modern evolutionary biology.** Contrary to Gould and Eldredge, I do not believe that the efforts of Simpson or any other paleontologist, however justly deserving of praise their efforts may have been, "brought paleontology within the modern synthesis of evolutionary theory" (1977, p. 115). Rather, I would claim that paleontologists abstracted and used those aspects of genetics and systematics that best fitted their preconceptions of the evolutionary process and, in fact, did not incorporate into their writings the most important theoretical construct the "New Systematics" had to offer, namely, the notion that species originate via allopatrically isolated populations after the instigation of some form of geographical isolation.

An examination of the theoretical postulates within paleontology and neontology over the past 50 years suggests that such a synthesis was philosophically and biologically difficult, if not impossible. The confusion and argumentation, I believe, support the view that the synthesis was more propaganda than a marriage pregnant with harmony. The primary reason for the lack of unification in paleontological and neontological evolutionary theory is the philosophical and biological debate over the reality of species as discrete evolutionary units. Paleontologists have maintained historically (1) that the reality of neontological species is illusionary; (2) that when projected back into evolutionary time, distinctions between "descendant species" (whether modern or not) and their "ancestral species" cannot be made; and therefore, (3) that, species are only arbitrarily designated segments of an evolutionary continuum. The quotes of Darwin and Simpson cited earlier are the historical extremes of a uniform body of paleontological opinion extending over 125 years, and the taxonomic literature is replete with additional examples. Paleontologists almost universally have adopted a nominalist philosophy. To accept gradualism *and* the reality of discrete species would be positively schizophrenic.[7]

Most neontologists, on the other hand, have viewed species as "real" biological entities, certainly discrete in space but ambiguously discrete in time.

---

[7] Without getting into an extended discussion, I note that some philosophers (e.g., Ruse, 1969) have presented arguments that species taxa, defined in terms of the biological species concept, are "real." Also, it can be mentioned that the species concept advocated here is consistent with the notion that species are to be viewed as individuals, not classes (Ghiselin, 1974; Hull, 1976). As far as I can see, the species concept of Simpson and many other paleontologists is totally inconsistent with this view.

With the time dimension ostensibly impenetrable to analysis, neontologists have said little about the unity of species in time, although they have been influenced by their paleontological colleagues. An important implication of the Eldredge and Gould model is that species can be viewed as discrete entities in geological time.

The battle against essentialism in the eighteenth and nineteenth centuries came to a climax with Darwin, and he opted (perhaps not consciously) for a nominalistic view of species, in part because it was impossible for him to visualize how evolution could proceed in discrete jumps. One suspects that the philosophy of the times would not permit such a view. Change was viewed as slow and gradual; therefore, natural selection must be slow and gradual. He confused, through no fault of his own, the issues of (a) selection acting on individuals of a species population and (b) the role of selection in the origin of species. Most paleontologists and many neontologists, it might be claimed, are still propounding this error.

Thus the adoption of the allopatric speciation model by paleontologists would establish a rapprochement with neontologists and should lead to a unified conceptual approach to common evolutionary problems.

**2. The allopatric speciation model will provide the conceptual foundation for an explosion of theory and methodology within paleontology.** We are beginning to see the effects of the model already, and in their joint papers, Eldredge and Gould have demonstrated the fresh insight that paleontology can look forward to in, for example, the study of morphology in space and time, adaptation, evolutionary rates, macroevolution theory, and numerous ecological phenomena such as character displacement (see Eldredge, 1974). Without meaning to impune the achievements of paleontologists in the past or the fine work of many contemporary paleontologists, phyletic gradualism is in a period of intellectual stasis; the period of development was about 30 to 40 years ago. *Even if the allopatric model is eventually proven incorrect and is replaced,* the benefits of its conceptual viewpoint for paleontology compared to the benefits of maintaining the status quo are obvious. This may be the best reason for its adoption.

**3. The allopatric speciation model will facilitate the incorporation within paleontology of a scientifically more rigorous methodology for reconstructing the history of life.** Adherence to a strict gradualistic viewpoint has canalized phylogenetic methodology within paleontology ("Search the rocks for ancestors, look for graduated links."). Such a research strategy has resulted in relatively little discussion of phylogenetic methodology by paleontologists; indeed, it would appear that most practicing paleontologists have felt little need to examine methodological principles (in fairness, the same criticism can be made of neontologists). The paleontological approach to phylogeny has often taken on the character of being essentially empirical (viewing fossils as directly historical data) rather than inferential.[8] Perhaps the most detrimental consequence of the gradualistic methodology and philosophy is that they have exempted paleontological thought from the notion that the history of life can be constructed in the absence of a fossil record, implying, if the fossil record is poor or nonexistent, that one should give up. I am convinced that this is a powerful reason why paleontologists and neontologists alike do not, in general, concern themselves with phylogeny. Even excluding those involved in purely stratigraphic studies, most paleontologists seem content to describe, identify, and name. Phylogenetic interpretation is wanting, as a perusal of the *Journal of Paleontology* will demonstrate.

The adoption of the allopatric speciation model will not be a panacea for these ills, but once paleontological minds are free of the dominion of gradualism, they will have an expanded capacity to consider other alternative methodologies for examining life's history.

## PUNCTUATED EQUILIBRIUM AND ALLOPATRIC SPECIATION: ARE THEY THE SAME?

Eldredge and Gould (1974, p. 304) claimed that they "lifted the notion of allopatric speciation *virgo intacta* from neontology, and have not produced [their] own variant of it. . . ." For the most part, this is true. I would like

---

[8] See the comments of Simpson (1960b, pp. 122–23) quoted at the end of this paper. Simpson is clearly not alone, for the paleontological literature is replete with statements to the effect that fossils are a requirement for reconstructing phylogeny or that we cannot know the phylogeny of a Recent group because the fossil record is inadequate, and so on. Implicit in many of these statements is the notion that only fossils are intrinsically capable of yielding historical data. I do not claim that this kind of thinking characterizes all of paleontology, only much of it.

to suggest that there is an important implication within the punctuated-equilibrium model that sets the latter somewhat apart from most current neontological thinking on speciation theory. The implication is derived from two assumptions of the punctuated-equilibrium model: that speciation can be considered to be a geologically instantaneous phenomenon and that morphological differentiation is concentrated near the time of isolation, with most species showing morphological stasis through time. If, for present purposes, these assumptions are accepted, then one conclusion must be that species can be considered discrete evolutionary entities *through geological time*. Neontologists themselves have been ambivalent on this matter and seemingly have been strongly influenced by the claims of paleontologists for phyletic gradualism. In general, neontologists have recognized the discreteness of species only at a given point in time (mainly the present). One neontological opinion, perhaps fairly typical, follows.

> The species, as manifested by a reproductive gap between populations, exists in full classical distinctness only in a local fauna. As soon as the dimensions of space (longitude and latitude) and time are added, the stage is set for incipient speciation. Populations will be found under these circumstances which are in the process of becoming separate species and have acquired some but not yet all of the attributes of distinct species (Mayr, 1969, p. 31).

> Species are evolving systems, and the vertical delimitation of species in the time dimension should in theory be impossible. . . . In most fossil sequences there are convenient breaks between horizons to permit a nonarbitrary delimitation of species (Mayr, 1969, p. 355).

It is not uncommon for neontologists to repeat the paleontological version of gradualism (e.g., Ross, 1974, pp. 58–60, and Mayr's comments above), although they clearly do not emphasize it. From the neontological viewpoint, allopatric speciation is often seen as a slow, gradual process (Mayr, 1970, p. 256). (Mayr's first quote above is thus reasonable for a narrow time scale, but not over geological time.) The punctuated-equilibrium model, because speciation is considered instantaneous in gological time, implies that species are evolutionary units whose integrity is maintained through time and which are reproductively isolated from other such units. Perhaps the concept of "methodological time" should be introduced in such discussions; in the vast majority of cases, our ability to resolve time in the fossil record only allows for "instantaneous" species formation.

# FURTHER IMPLICATIONS OF PUNCTUATED EQUILIBRIUM

The punctuated-equilibrium model is a powerful, heuristic concept for paleontology. Speciation by geographically isolated populations is by far the most common evolutionary mode in sexually reproducing organisms (Bush, 1975). Clearly, other modes of speciation are operable in nature, but one can reasonably assert, as a general rule, that allopatric speciation is applicable to the vast majority of organisms with a fossil record good enough to investigate problems of evolutionary mode.

With the punctuated-equilibrium model as a framework, paleontologists have an opportunity to make significant contributions to the study of evolutionary modes. The following questions arise: How fast is "geologically instantaneous"? Do species remain relatively morphologically static in time? Is intraspecific variability through space and at one time interval more or less than the interspecific variability of parent–daughter species at the time of speciation? What are the morphological and distributional consequences of sympatry through time? Do patterns of speciation vary from group to group?

It will take some time for many paleontologists to realize fully the importance of punctuated equilibrium as a conceptual model. Some, for example, consider it to be "quantum evolution of species in marginal populations and their subsequent expansion into the fossil record" (Simpson, 1976, p. 5). Punctuated equilibrium and the "quantum evolution" of Simpson (1944, 1953) are very distinct concepts. The latter is a "special, more or less extreme and limiting case of phyletic evolution" (Simpson, 1953, p. 389) and involves rapid evolution, under strong directional selection, of a population from one "adaptive zone" to another. There are discontinuities between "zones" in this model, and transitional forms between zones are not believed to persist. This scenario is not punctuated equilibrium (or allopatric speciation), and indeed, the punctuated-equilibrium model may make quantum evolution superfluous.

As should be evident to any working paleontologist, it will require a very good fossil record, spatially and temporally, to test and study the punctuated-equilibrium model. Gingerich (1976, pp. 22–24) has correctly pointed out that it is not useful to argue for punctuated equilibrium in cases in which there is inadequate stratigraphic control. The major problem with examples of gradualism is the lack of lateral control, and lateral control will be particularly important in the analysis of punctuated equilibrium since speciation is first and foremost a spatial phenomenon.

It should be obvious at this point that punctuated equilibrium offers paleontology a radical alternative to gradualism—radical in the sense of a distinctly different conceptual framework. Recently, Eldredge (1979) has emphasized this distinction by showing that an adherence to a gradualistic philosophy (called the transformational approach) has clouded our perceptions of the evolutionary process in rather fundamental ways. Questions have been framed in terms of the transformation of features (gene content, morphology, behavior) from one state to another, and the central theme of the morphologic postulate is adaptation, accomplished through natural selection (Eldredge, 1979).

But there is an alternative for evolutionary biology, and that is the taxic approach. Here the major question concerns the origin of taxa (species); changes in genes, morphology, behavior, and so forth, are to be interpreted within the context of taxic evolution. What Eldredge and Gould have been saying, and what I have been emphasizing here, is that the taxic approach has the potential to effect a major change in evolutionary paleontology.

## Reconstructing the History of Life

Paleontologists and neontologists have repeatedly recognized the inability of the fossil record to yield Darwinian phylogenies in which forms intergrade smoothly with one another through time. There are discontinuities—temporal, spatial, and morphological—among specimens; paleontologists, like neontologists, have traditionally used these gaps as the basis for inferring taxonomic distinction for samples of similar specimens. It is of little importance for present purposes whether we look upon these taxonomic entities as "unit taxa" or specify them as being "species" on the supposition that they exemplify, in some way, one's preferred species concept.[9] I will call these entities species. We now must ask the fundamental question: How do we reconstruct the phylogenetic history of these species? By phylogenetic history I mean the results of specifying the relationships (of whatever kind) among the species in question. Historically, the paleontologist studying the fossil record from a conceptual viewpoint of gradualism and nondiscrete species has been compelled to search the rocks for ancestors and descendants; indeed, with this viewpoint the stratigraphic record would seem to be the only recourse because the item of interest *is* the gradual transition. If, however, species are viewed as geologically and biologically discrete, then interest is shifted from the transition per se to hypotheses about relationships among these discrete entities.

## THE CONFLICT

The conflict within neontology and paleontology over the research strategy to be used in phylogenetic analysis focuses on one major issue—the kinds of relationship to be specified (see Cracraft, 1974, for additional discussion). There are two kinds of relationships among species: ancestor–descendant relationships and relationships by common ancestry. Ancestor–descendant relationships are derived directly from the Darwinian gradualistic philosophy; common ancestry also comes from Darwin and pertains to the splitting of an ancestral population to produce two species where there once was one. Most of the argument has been about whether ancestors should, and can, be specified, the relative importance of morphological and temporal data, and whether phylogenetic analysis is amenable to hypothetico-deductive procedures.

The schools of thought regarding phylogenetic methodology that have

---

[9] These "unit taxa" are collections of specimens that have variability in space and time. In phylogenetic analysis, we are interested in the hierarchical arrangement of these taxa; their variability is important only in so far as it bears on the identification and delimitation of the taxa in the first place. Phylogenetic analysis is concerned with phenomena in terms of the taxic approach (Eldredge, 1979).

become defined in the last decade are diverse in opinion, and it would do each a disservice to attempt a simple characterization. The so-called evolutionary systematists—Simpson (1961), Mayr (1969), and Bock (1973)—have much in common but differ in many of their viewpoints (Nelson, 1972). The school of numerical taxonomy (Sneath and Sokal, 1973), although not phylogenetic in outlook, has produced ideas and methodology that are being incorporated by some recent paleontological workers whose philosophy is close to that of the evolutionary systematists (Bretsky, 1971, and in this volume; Gingerich, 1974, 1976, and in this volume). The literature and theory of phylogenetic systematics (Hennig, 1966) has expanded enormously in the last 10 years, and a completely uniform body of opinion cannot be said to exist, although there is substantial agreement on philosophy and methodology (see papers by Gaffney and Eldredge, in this volume, for a summary).

This diversity of theory and methodology in phylogenetic strategies would seem to give little hope for rapprochement, and positions remain polarized. It is my opinion, however, that the issues basic to a methodology of phylogeny reconstruction are relatively simple, and it is the purpose of what follows to outline them briefly. Needless to say, my conclusions and interpretations may be unacceptable to some for, quite frankly, they support the school of phylogenetic systematics. Nevertheless, in the following treatment of these issues I will attempt to point to major areas of disagreement and seek some resolution. I will not be concerned with a detailed presentation of methodology, since that will be covered by others in the symposium.

## THE HIERARCHICAL STRUCTURE OF NATURE

It has long been accepted that the phylogenetic history of organisms has a geometry of branching and divergence. That there is branching implies a hierarchical structure to that history. Perhaps the first fundamental problem of systematics is to discover, to the extent possible, the form of that hierarchical structure. *Cladograms* are hypotheses about the structure of nature's hierarchy.

The word cladogram is used here in a sense slightly different from that of most systematists. A cladogram is literally a hierarchical branching diagram that says something about the shared similarities of the taxa included in that diagram. It does not represent a phylogeny of these taxa—that is the purpose of a *tree*. A number of trees might be consistent with a given cladogram since the latter says nothing about the nature of the relationships among taxa, only their shared similarities. Trees, on the other hand, include assumptions about the two kinds of relationships.

What kinds of shared similarities are represented by a cladogram? There are only two: nonevolutionary (nonhomologous, or convergent) and evolutionary (homologous). Two taxa, say A and B, can share two kinds of evolutionary similarities. First, they may share similarities that were inherited from ancestors more distant than their immediate common ancestor; these are shared *primitive similarities* (the "symplesiomorphies" of Hennig, 1966). Second, they may share similarities inherited only from their immediate common ancestor and not shared with more distantly related taxa; these are shared *derived similarities* (the "synapomorphies" of Hennig). A cladogram is, therefore, a branching diagram that depicts the nested patterns of shared evolutionary similarities (homologies) of all the taxa under study, and the hypothesis is that the patterns reflect the hierarchical structure of the taxa themselves. Homologous features, whether morphological, behavioral, physiological, or biochemical, can always be considered derived at one level of the hierarchy and primitive at all lower levels. Hence the structure of the hierarchy is defined in terms of shared derived similarities (synapomorphies) since they are the only evolutionary similarities capable of defining sets of taxa within the hierarchy.[10]

It should be evident that cladograms, as conceived of here, have no inherent temporal content and are constructed by comparative analysis of the attributes possessed by the taxa in question. Thus stratigraphic data are irrelevant for cladogram construction. Taxa A and B may be considered end points of a simple branching diagram because of their shared derived similarities, even though we have not yet considered the alternative propositions that (1) A and B are derived from a common ancestor, (2) A is the ancestor of B, or

---

[10] This is an important point. Branching and divergence produce sets of nested taxa. What kinds of set-defining properties (similarities) are consistent with currently accepted evolutionary theory? To my knowledge, theory implies that the only kinds of shared similarities capable of defining sets of taxa are evolutionary novelties (i.e., derived similarities). This does not mean, of course, that a systematist may not choose to use some other concept of similarity—say, the general *overall* similarity of pheneticists—but evolutionary theory does not imply that nested sets of taxa can be defined using that concept.

(3) B is the ancestor of A. These propositions lie within the domain of tree analysis.

## TREE ANALYSIS

Neontologists and paleontologists are interested primarily in the history of life, that is, phylogeny, and thus are more concerned with trees than cladograms. But it must be stressed, in the strongest terms possible, that the process of choosing a particular cladogram should precede any decision about which tree, of the ones that are consistent with the cladogram, is to be preferred. As will be emphasized by others in this symposium, much of the argumentation within systematics has been unnecessary because it has been directed toward controversies over trees rather than cladograms.

The central problem of tree analysis is deciding whether each branch point of a cladogram is to be retained or eliminated. If the branch point is retained, then it is hypothesized that the branches from that point are the result of a speciation event, and corresponding taxa are considered related by common ancestry, with no ancestral taxon being specified. If the branch point is eliminated, then it is hypothesized that the species of one branch is the ancestor of the taxon of the coordinate branch. With respect to specifying an ancestor, it should be obvious that only a branch with one species can be considered to be an ancestor; if a branch possesses another, more distal branch, then the main branch cannot be specified as an ancestor (in other words, a supraspecific taxon cannot be the ancestor of another species or supraspecific taxon).[11]

Over the years, many phylogenetic systematists, myself included, have advocated the adoption of phylogenies in which ancestors are not specified (Cracraft, 1974). In more formal terms, what we have been saying is that

---

[11] This statement may invite the comment of splitting hairs. However, its adoption assures two salutary results: the elimination of bad biology and the implementation of precision in the analysis of historical events. Single species give rise to species; speciation theory does not support the recognition of ancestral supraspecific taxa. All this implies is that tree analysis is most properly performed at the species level.

these trees should retain all the branch points of the cladogram on which they were based [such trees might be called X-trees (see Eldredge, in this volume); Nelson, personal communication]. Much of the argumentation about this issue results from the opponents speaking a different philosophical language. Perhaps the crucial issue in phylogenetic analysis is, Are we going to agree that hypotheses must be formulated such that they can be tested in a rigorous and consistent manner, or are we willing to accept the kinds of explanations whereby each investigator can have his own personal picture, a picture which is essentially untestable because of *ad hoc* assumptions? The problem of hypothesis formulation and testing is central to the scientific efficacy of both cladograms and trees. It is my opinion that the second type of explanation, noted above, has no heuristic or scientific value with regard to cladograms or trees.

## FORMULATION AND TESTING OF HYPOTHESES ABOUT CLADOGRAMS AND TREES

The principles of formulating and testing cladograms and trees are relatively simple; the practical application of these principles to a particular group of organisms may be easy, difficult, or impossible. This should not be surprising, yet some critics attack phylogenetic systematics for what is termed its "simplistic" theoretical and methodological approach to nature's complexity. Within the philosophy of science, the most simple theory structure is to be preferred as long as explanatory power is not compromised. To my knowledge, no one has challenged the explanatory power of phylogenetic systematics, and I suggest that if any "information" is being excluded, it is of the untestable type. No phylogenetic systematist would claim that all phylogenetic analysis is simple, but all would claim, I believe, that there is precious little room for art in the science of systematics.

The application of hypothetico-deductive reasoning to the evaluation (testing) of cladograms has been discussed in considerable detail by Wiley (1975, 1976) and will be treated extensively by Gaffney (in this volume). Stated simply, postulated synapomorphies are used to falsify alternative cladograms, and the cladogram with the fewest falsifications is to be preferred.[12] The choice among alternative cladograms is thus based on the criterion of parsimony, "not because nature is parsimonious, but because only parsimonious hypotheses can be defended by the investigator without resorting to authoritarianism or apriorism" (Wiley, 1975, p. 236). Another way of viewing cladogram testing is to say that we prefer the cladogram that exhibits the fewest convergences; in this sense, convergences are postulated to be nonevolutionary similarities (nonhomologies). It should be evident that a similarity postulated as being a shared derived similarity in one cladogram could be postulated as a convergence in another cladogram. Finally, the more postulated synapomorphies that are consistent with a particular cladogram, the more we can view that cladogram as being corroborated and alternative cladograms as being falsified.

In testing phylogenetic trees, we are faced with much the same strategy as in testing cladograms. Indeed, if the trees are X-trees, that is, if all relationships are assumed to be of common ancestry, then the procedures are identical. There are differences, however, when ancestor–descendant relationships are postulated (see the detailed discussions of Engelmann and Wiley, 1977, Platnick, 1977; paleontologists with opposing viewpoints include Harper, 1976; Szalay, 1977). Given the hypothesis that taxon A is the ancestor of taxon B, there are two tests. First, if either A or B shares a derived similarity with some third taxon, the ancestor–descendant relationship is refuted. Second, if taxon A is found to possess a unique, derived character state (*autapomorphy*) that is not found in taxon B or any other taxon (i.e., the other taxa possess the primitive condition relative to A), the hypothesis can be said to be refuted. The only character-state distribution that is consistent with the hypothesis is that taxon A possesses character states that are all primitive relative to those in B (unless, of course, they are shared synapomorphies relative to a condition in a third taxon). This second test is unsatisfactory for rigorously evaluating the hypothesis because (1) it is also consistent with the hypothesis that A and B share a common ancestor and

---

[12] The word "falsify" is not used here in a strict sense, i.e., to prove false, for clearly no hypothesis of relationships could be *proven* true or false. Rather, in phylogenetic analysis, an hypothesis of relationships can be said to be falsified if a perceived synapomorphy is *inconsistent* or *in conflict* with the hypothesis. In this sense, phylogenetic hypotheses are comparable to scientific hypotheses in general; it is rare for a hypothesis, even the most widely accepted, not to be in conflict with some empirical observations.

(2) it is based on negative evidence since it is always possible that a unique feature in A has gone undetected. Platnick (1977) concludes, and I find his argument compelling, that ancestor–descendant relationships as expressed in trees are not scientifically testable in any definitive sense: a hypothesis of an ancestor–descendant relationship can never be the sole hypothesis left unrefuted (one cannot, at the same time, eliminate the hypothesis of common ancestry from some unknown ancestor).

# The Principles of Evolutionary Paleontology and Neontology: Toward a Theory of Comparative Biology

At the present time, evolutionary paleontology operates within a loose set of guidelines governed by tradition rather than a concisely formulated body of theory. It is a tradition defined by the belief that the key to life's history lies more in empirical observation than in inference. Systematics in general, and phylogenetic analysis in particular, has often been viewed as art rather than science, or at least as artistic science. It is also clear that, with respect to theory and method, paleontology has been isolated from neontology. Neontologists have traditionally given to paleontology the responsibility of interpreting life's history, with the misunderstanding, on both sides, that the dimension of time is somehow the key to that history. As Simpson (1960a, p. 9) expressed it, "The processes of life can be adequately displayed only in the course of life throughout the long ages of its existence." No one would deny that history implies a time dimension, whether relative or absolute, but I suggest that the overemphasis on time by neontologists and paleontologists has obstructed the development of a systematic approach to the study of life's history and a unified theory of comparative biology applicable to neontology and paleontology alike. This tension between the neontological and paleontological viewpoints is perhaps nowhere better stated than by Simpson himself:

> The great drawback of the comparative method and of contemporaneous evidence is that they are not in themselves historical in nature. The drawing of historical conclusions from them is therefore full of pitfalls unless it can be adequately con-

trolled by *directly* historical evidence . . . , [which] either involves lengths of time that are, to the historian of life, infinitesimally short or is drawn from the preserved parts of organisms long dead" (1960, pp. 122–23).

I do not believe that I am exaggerating the theoretical and methodological schism between neontology and paleontology. In their classic works on systematics, Simpson (1961) operated almost wholly within a paleontological perspective, Mayr (1969) within a neontological one. Part of this is their background, surely, but the divisions are more fundamental, as I have noted throughout this paper. As a consequence, it cannot be claimed that their writings have produced a general method of comparative biology.

The introduction of phylogenetic systematics into neontology began 25 years ago (Hennig, 1950), but it has been only in the last 10 years that a significant impact from its theory and method has been felt (Hennig, 1966). Without question, there has been a profound effect on neontological systematics, and it is inevitable that this influence will continue to expand. Introduction of this theory and method into paleontology occurred less than 5 years ago, and an increasing number of paleontologists are adopting its principles. The reason for this fundamental change in thinking by so many workers is not difficult to understand.

For the first time, a general theory of comparative biology is being forged (Nelson, 1970). It has as its foundation the proposition that the complexity of life's history can be studied by a scientific, comparative method and that this method is hypothetico-deductive in structure. Precise hypotheses can be formulated and subjected to rigorous testing procedures. That this scientific framework has been lacking within systematics and paleontology is painfully obvious.

To be sure, the theory of phylogenetic systematics continues to change, but the core of its comparative method seems secure. The theory and method have demonstrated a generality applicable to phylogenetic analysis, classification theory, and biogeography—in short, precisely those branches of biology that are comparative and historical in their content.

It is, I believe, a safe prediction that the theory of phylogenetic systematics will have as profound an effect on the methodology of evolutionary paleontology as the introduction of the allopatric speciation model. Indeed, they are highly compatible with each other, and it is not simply a matter of coincidence that they have been generating considerable interest within paleontology at about the same time. Both will produce, despite considerable skepticism at first, notable progress in the future.

## Acknowledgments

I want to thank David L. Hull for his consummate comments on the manuscript. Niles Eldredge provided discussions of constant good wit and advice. I thank the National Science Foundation (DEB76-09661) for their support while I was writing this paper.

## References

Bock, W. J. 1973. Philosophical foundations of classical evolutionary classification. *Syst. Zool.* 22:375–92.
Bretsky, S. S. 1971. Evaluation of the efficacy of numerical taxonomic methods: an example from the bivalve mollusks. *Syst. Zool.* 20:204–22.
Burkhardt, R. W., Jr. 1977. *The Spirit of System: Lamarck and Evolutionary Biology.* Harvard University Press, Cambridge, Mass.
Bush, G. L. 1975. Modes of animal speciation. *Ann. Rev. Ecol. Syst.* 6:339–64.
Butterfield, H. 1957. *The Origin of Modern Science.* Collier Books, New York.
Coleman, W. 1962. Lyell and the "reality" of species: 1830–33. *Isis* 53:325–38.
Coleman, W. 1964. *Georges Cuvier, Zoologist.* Harvard University Press, Cambridge, Mass.
Cracraft, J. 1974. Phylogenetic models and classification. *Syst. Zool.* 23:71–91.
Darwin, C. 1859. *On the Origin of Species,* facsimile of 1st ed. Atheneum, New York, 1967.
Darwin, C. 1872. *On the Origin of Species,* 6th ed. Doubleday, New York.
Dobzhansky, T. 1937. *Genetics and the Origin of Species.* Columbia University Press, New York.
Eldredge, N. 1971. The allopatric model and phylogeny in Paleozoic invertebrates. *Evolution* 25:156–67.
Eldredge, N. 1972. Systematics and evolution of *Phacops rana* (Green, 1832) and *Phacops iowensis* Delo, 1935 (Trilobita) in the middle Devonian of North America. *Bull. Amer. Mus. Nat. Hist.* 47:45–114.
Eldredge, N. 1974. Character displacement in evolutionary time. *Amer. Zool.* 14:1083–97.
Eldredge, N. 1979. Alternative approaches to evolutionary theory. In J. H. Schwartz and H. B. Rollins (eds.), *Models and Methodologies in Evolutionary Theory,* vol. 1. Carnegie Museum Special Publication. (In press)

Eldredge, N., and Gould, S. J. 1972. Punctuated equilibrium: An alternative to phyletic gradualism. In T. J. M. Schopf (ed.), *Models in Paleobiology* pp. 82–115 Freeman, Cooper & Co. San Francisco.

Eldredge, N., and Gould, S. J. 1974. Reply to M. K. Hecht (1974). *Evol. Biol.* 7:303–8.

Engelmann, G. F., and Wiley, E. O. 1977. The place of ancestor–descendant relationships in phylogeny reconstruction. *Syst. Zool.* 26:1–11.

Gale, B. G. 1972. Darwin and the concept of a struggle for existence: A study in the extrascientific origins of scientific ideas. *Isis* 63:321–44.

Ghiselin, M. T. 1974. A radical solution to the species problem. *Syst. Zool.* 23:536–44.

Gingerich, P. D. 1974. Stratigraphic record of early Eocene *Hyopsodus* and the geometry of mammalian phylogeny. *Nature (London)* 248:107–9.

Gingerich, P. D. 1976. Paleontology and phylogeny: Patterns of evolution at the species level in early Tertiary mammals. *Amer. J. Sci.* 276:1–28.

Gould, S. J., and Eldredge, N. 1977. Punctuated equilbria: The tempo and mode of evolution reconsidered. *Paleobiology* 3:115–51.

Greene, J. C. 1961. *The Death of Adam.* New American Library, New York.

Harper, C. W., Jr. 1976. Phylogenetic inference in paleontology. *J. Paleontol.* 50:180–93.

Hecht, M. K. 1974. Morphological transformation, the fossil record, and the mechanisms of evolution: A debate. *Evol. Biol.* 7:295–303.

Hennig, W. 1950. *Grundzüge einer Theorie der phylogenetischen Systematik.* Deutscher Zentralverlag, Berlin.

Hennig, W. 1966. *Phylogenetic Systematics.* University of Illinois Press, Urbana.

Hull, D. L. 1965. The effect of essentialism on taxonomy—two thousand years of stasis. *Brit. J. Phil. Sci.* 15:2–14; ibid., 16:15–32.

Hull, D. L. 1967. The metaphysics of evolution. *Brit. J. Hist. Sci.* 3:309–37.

Hull, D. L. 1976. Are species really individuals? *Syst. Zool.* 25:174–91.

Huxley, J. S., ed. 1940. *The New Systematics* Oxford University Press, London.

Huxley, J. S. 1942. *Evolution, the Modern Synthesis.* George Allen & Unwin, London.

Lamarck, J. B. 1914. *Zoological Philosophy,* H. Elliot, trans.; originally pub. 1809. Macmillan & Co., New York.

Lovejoy, A. O. 1936. *The Great Chain of Being.* Harvard University Press, Cambridge, Mass.

Lovejoy, A. O. 1959. The argument for organic evolution before "The Origin of Species," 1830–1858. In B. Glass, O. Temkin, and W. L. Straus, Jr. (eds.), *Forerunners of Darwin: 1745–1859,* pp. 356–414. Johns Hopkins Press, Baltimore.

Mayr, E. 1942. *Systematics and the Origin of Species.* Columbia University Press, New York.

Mayr, E. 1963. *Animal Species and Evolution.* Harvard University Press, Cambridge, Mass.

Mayr, E. 1969. *Principles of Systematic Zoology.* McGraw-Hill, New York.

Mayr, E. 1970. *Populations, Species, and Evolution.* Harvard University Press, Cambridge, Mass.

Mayr, E. 1972. The nature of the Darwinian revolution. *Science* 176:981–89.
Nelson, G. J. 1970. Outline of a theory of comparative biology. *Syst. Zool.* 19:373–84.
Nelson, G. J. 1972. Comments on Hennig's "Phylogenetic Systematics" and its influence on ichthyology. *Syst. Zool.* 21:364–74.
Osborn, H. F. 1910. Evolution as it appears to the paleontologist. Paper presented at the 7th International Zoological Congress. Boston, 1907.
Platnick, N. I. 1977. Cladograms, phylogenetic trees, and hypothesis testing. *Syst. Zool.* 26:438–42.
Ross, H. H. 1974. *Biological Systematics*. Addison-Wesley, Reading, Mass.
Ruse, M. 1969. Definitions of species in biology. *Brit. J. Phil. Sci.* 20:97–119.
Ruse, M. 1970. The revolution in biology. *Theoria* 36:1–22.
Simpson, G. G. 1944. *Tempo and Mode in Evolution*. Columbia University Press, New York.
Simpson, G. G. 1953. *The Major Features of Evolution*. Columbia University Press, New York.
Simpson, G. G. 1960a. *The Meaning of Evolution*. Yale University Press, New Haven, Conn.
Simpson, G. G. 1960b. The history of life. In S. Tax (ed.), *Evolution of Life*, vol. 1. University of Chicago Press, Chicago.
Simpson, G. G. 1961. *Principles of Animal Taxonomy*. Columbia University Press, New York.
Simpson, G. G. 1976. The compleat palaeontologist? *Ann. Rev. Earth Planet. Sci.* 4:1–13.
Sloan, P. R. 1976. The Buffon–Linnaeus controversy. *Isis* 67:356–75.
Sneath, P. H. A., and Sokal, R. R. 1973. *Numerical Taxonomy*. W. H. Freeman & Co., San Francisco.
Sylvester-Bradley, P. C., ed. 1956. *The Species Concept in Paleontology*. London Systematics Assoc. Publ. no. 2.
Szalay, F. S. 1977. Ancestors, descendants, sister-groups and testing of phylogenetic hypothesis. *Syst. Zool.* 26:12–18.
Toulmin, S., and Goodfield, Jr. 1977. *The Discovery of Time* (Midway reprint, orig. pub. 1965). University of Chicago Press, Chicago.
Vorzimmer, P. J. 1970. *Charles Darwin: The Years of Controversy*. Temple University Press, Philadelphia.
Wiley, E. O. 1975. Karl R. Popper, systematics, and classification: A reply to Walter Bock and other evolutionary taxonomists. *Syst. Zool.* 24:233–43.
Wiley, E. O. 1976. The phylogeny and biogeography of fossil and Recent gars (Actinopterygii: Lepososteidae). *Univ. Kans. Mus. Nat. Hist. Misc. Publ.* 64:1–111.
Zittel, K. A. von. 1913. *Text-book of Paleontology* (orig. ed. 1899; C. R. Eastman, ed.). Macmillan & Co., London.

# The Stratophenetic Approach to Phylogeny Reconstruction in Vertebrate Paleontology

## Philip D. Gingerich

Phylogeny and methods of reconstructing phylogeny are basic to our understanding of evolutionary history in the vertebrates, as in any other group of organisms. I have been asked to present an "evolutionary" approach to phylogeny reconstruction in vertebrate paleontology. The evolutionary approach is usually associated with Professors George Gaylord Simpson and Ernst Mayr, both of whom have contributed greatly to our understanding of vertebrate phylogeny and evolution. Simpson (1975, 1976) and Mayr (1974) have recently summarized their views on phylogeny and phylogeny reconstruction. In this paper, I would like to take a slightly different approach from that presented by others and emphasize the importance of a dense and continuous fossil record for understanding the phylogeny of vertebrates. After considering the meaning and importance of phylogeny, I shall outline what I have called the "stratophenetic" approach to phylogeny reconstruction, applying it first to the evolution of mammals at the species level and then to more general problems of vertebrate phylogeny.

# Phylogeny and Evolutionary Theory

Phylogeny is usually understood to mean the evolutionary history of a group of organisms. As such, a phylogeny includes genealogical information on the interrelationships of all the individual members of the group. This pedigree (from Middle French *pié de grue,* or "crane's foot") is normally expressed in a dendrogram. The adaptations of members of a group of organisms are equal in importance to the group's pedigree for understanding its evolutionary history. Pedigree and adaptation together describe evolution, and it remains only to stress the importance of time as the principal dimension of history. In brief, phylogeny as evolutionary history traces both genealogical relationships and adaptations through time.

Time sufficient for the study of evolutionary history is a unique attribute of geology. Hence it is not surprising that the establishment of evolution as a historical fact was made by geologists and paleontologists. Hutton's original essay on the "Theory of the Earth" (1788) found "no vestige of a beginning, no prospect of an end" to geological time, infinitely expanding earlier estimates of earth history. Lamarck, Cuvier, and Brongniart in France and Smith, Lyell, and Darwin in England all had firsthand experience working with fossils and their distribution in geological formations. The principle of faunal succession that we attribute to William Smith is an empirical principle that lies at the heart of our understanding of evolution. In a very real sense, evolution is a geological phenomenon.

The empirical nineteenth-century paleontological observation that life evolved through geological time required a biological explanation and, in this century, the study of evolution has become predominantly a biological subject. As Darwin first postulated, biological variation, inheritance, overpopulation, and differential reproduction can be combined into one hypothesis, in which natural selection leads to marked and important evolutionary changes in plant and animal species. Enormous advances have been made in our understanding of each of the major components of evolutionary theory since the publication of Darwin's *Origin of Species* more than a century ago, and we are now at a point where paleontology is once again regaining its former importance in the study of evolution. Many hypotheses of tempo and mode in speciation, and rate and direction in phylogeny, can only be tested by the empirical evidence of paleontology. The fossil record is much better known than it was in Darwin's time, and we now see this wealth of paleontological data generating new hypotheses as well as testing older ones.

## SPECIES

The basic unit of evolution is the species. Species may be defined in various ways—one of the most widely used modern definitions being Mayr's (1970): "species are groups of interbreeding natural populations that are reproductively isolated from other such groups." Operationally, living species are almost always recognized as phenetic clusters of very similar individuals (Sokal and Crovello, 1970) since data are usually not available to support inferences about interbreeding or reproductive isolation. Morphological variation is distributed continuously within such phenetic clusters, but nonarbitrary discontinuities separate species clusters. Viewing species as phenetic clusters within a multivariate morphometric space emphasizes both the nonarbitrary nature of species and their relational definition since a given multivariate cluster only exists relative to another cluster or clusters. When geographic variation or temporal variation is included, the objectivity (but not the reality) of a species is sometimes reduced, and here again, continuity in the distribution of morphological, geographical, and temporal characteristics is necessary to unify each species.

Taking a simplified example of the relational definition of species as phenetic clusters, we can look at a hypothetical *"Canis"* morphological axis in a multivariate morphometric space centered over the Upper Peninsula of Michigan during the last century or so (an instant in geological time). Leaving out domestic dogs, we very clearly find two wild species: the wolf and the coyote. It is sometimes argued that wolves and coyotes are really the same species since they occasionally interbreed and either of them can interbreed with feral dogs, and since wolves sometimes appear to displace coyotes ecologically. The fact remains that both are found in the same gen-

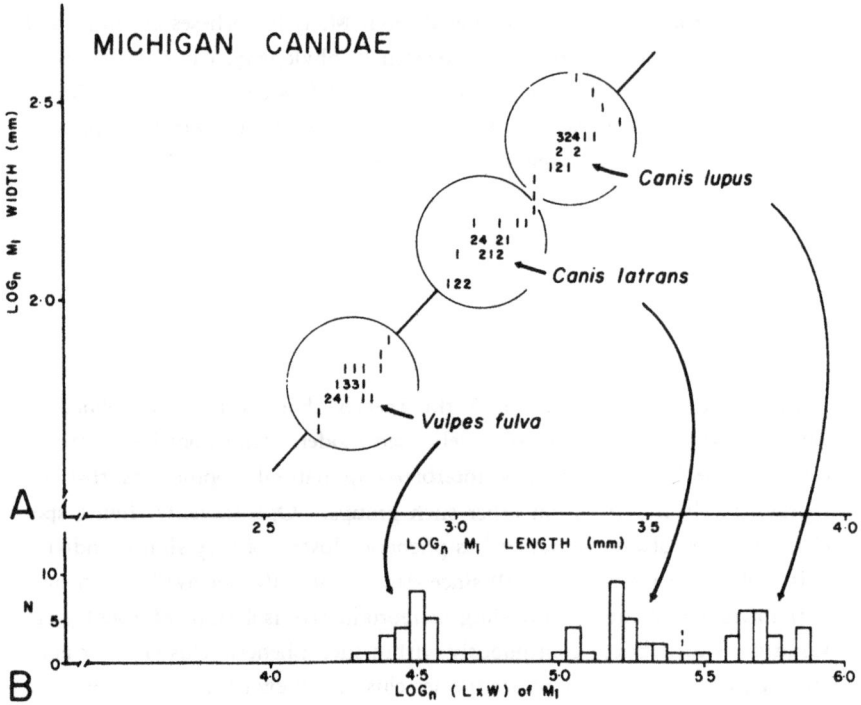

Figure 1. Distribution of tooth size in three sympatric species of living Canidae from the Upper Peninsula of Michigan. (A) Scatter plot of log length vs. log width of $M_1$, the lower carnassial tooth, for red foxes (*Vulpes fulva*), coyotes (*Canis latrans*), and wolves (*Canis lupus*). (B) Histogram of the log of length multiplied by width of $M_1$, showing the same pattern of grouping as the principal axis of the scatter plot. Note that each species occupies 0.45 units on the logarithmic horizontal scale of the histogram, and note also the contiguity of distributions for coyotes and wolves. All specimens are in the University of Michigan Museum of Zoology. See Rosenzweig (1966) for an ecological interpretation of size differences in sympatric carnivores.

eral geographic areas and have the same general morphology, but the two differ significantly in size. Plotting the size of their carnassial teeth (fig. 1), or any other measure of body size, we see that wolves clearly form a cluster at a larger overall size than the otherwise similar coyote cluster.

The fundamental difference between wolves and coyotes is a difference in body size. Paleontologists sometimes disparage size differences between taxa, stating that a "real" difference would be reflected in significant differences in form. Considering the role that allometry plays in maintaining functional equivalence during morphogenesis, many minor differences in form can be adequately explained as a simple result of size differences. Furthermore, there is an increasingly large literature in ecology on the importance of body size as one of the most fundamental components of an animal's adaptation (Hutchinson and MacArthur, 1959; McNab, 1971; Stanley, 1973; Van Valen, 1973). Thus, not only do wolves and coyotes form two distinct phenetic clusters, but the major difference between the two clusters (overall body size) is probably the most important adaptive difference between the two. Tooth size is very highly correlated with body size in mammals ($r = 0.93$–$0.98$, Gould, 1975), and the teeth on which mammalian paleontologists base so much of their knowledge offer a direct pathway to determining an animal's body size (see fig. 2). I shall use this relation of tooth size to body size to infer that significant evolutionary changes in tooth size reflect important changes in overall body size and related adaptations.

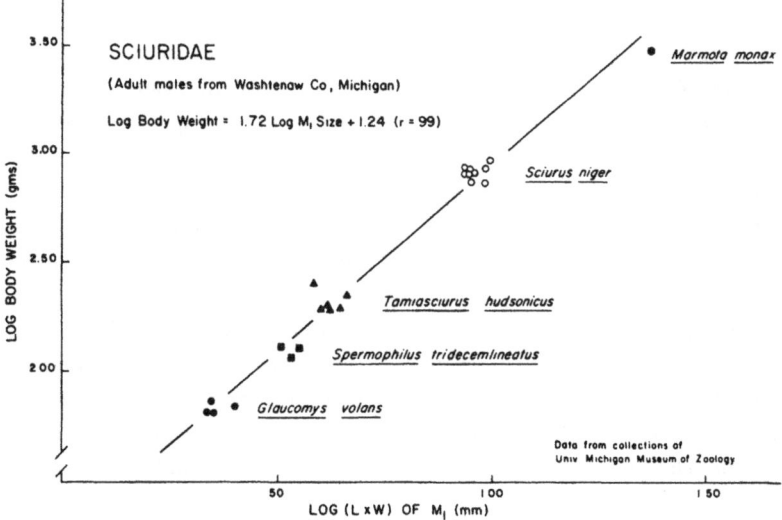

Figure 2. Correlation of body weight with tooth size in five sympatric species of squirrels of the family Sciuridae, ranging from the flying squirrel (*Glaucomys*) to the marmot (*Marmota*). A similarly high correlation obtains over the entire range of generalized mammals, which makes tooth size a good predictor of the average body size of mammalian species. This correlation can be used in the fossil record to compare the relative body size of related genera and species. Figure is from Gingerich (1976b).

## LINEAGES

Thus far we have considered species as relationally defined phenetic clusters at one point in space and time. If we trace one phenetic cluster (one species) over its geographic range, it often changes, hence the familiar geographic gradients that Huxley termed "clines." In one of the few attempts to quantify clinal variation in body size, McNab (1971) found relatively minor changes with latitude. Much steeper clinal variation is seen in temporal changes in some species. Simpson (1943) termed these temporal clines "chronoclines" to distinguish them from geographic clines. Subsequently, chronoclines have been documented by other authors, one of the best examples being furnished by the early Eocene primate *Pelycodus* (fig. 3).

If we examine the collection of mammals from one place and time, for example, the Big Horn Basin of Wyoming at level 160 (shown in fig. 3), we find several specimens of a single small phenetic cluster placed in the genus *Pelycodus*. Two specimens have first lower molars ($M_1$) that can be measured, and these are 12.8 and 12.9 mm² in crown area. The specimens are very nearly the same size, and plotted on a log scale, they fall very close together, at the bottom of the sequence in figure 3. By examining the collection of mammals from a slightly later time, level 180, specimens of a single phenetic cluster of *Pelycodus* are again found. Here again, two first lower molars can be measured and they yield measurements of 12.9 and 14.4 mm². Fifteen specimens preserve $M_1$ in the single phenetic cluster of *Pelycodus* from level 200, and these range from 10.7 to 15.1 mm². When all of the specimens of *Pelycodus* from lower Graybull beds that preserve $M_1$ from about level 200 to level 540 are plotted, a distinctive pattern emerges. Allowing for some relatively minor variation, there is a continuous and gradual change

Figure 3. Detailed stratigraphic distribution of primate *Pelycodus* and the related genus *Copelemur* in early Eocene sediments of the Big Horn Basin, Wyoming. Abscissa is tooth size and, by inference, relative body size. (Stratigraphy is discussed in Gingerich, 1976a.) Numbers at left are stratigraphic position (in feet) above the base of the Willwood Formation. Horizontal line is sample range; vertical slash is mean; solid bar is standard error of mean; small number at right is the sample size. Dashed lines show the pattern of stratophenetic linking for species of North American *Pelycodus* (compare with fig. 8). Figure is from Gingerich and Simons (1977).

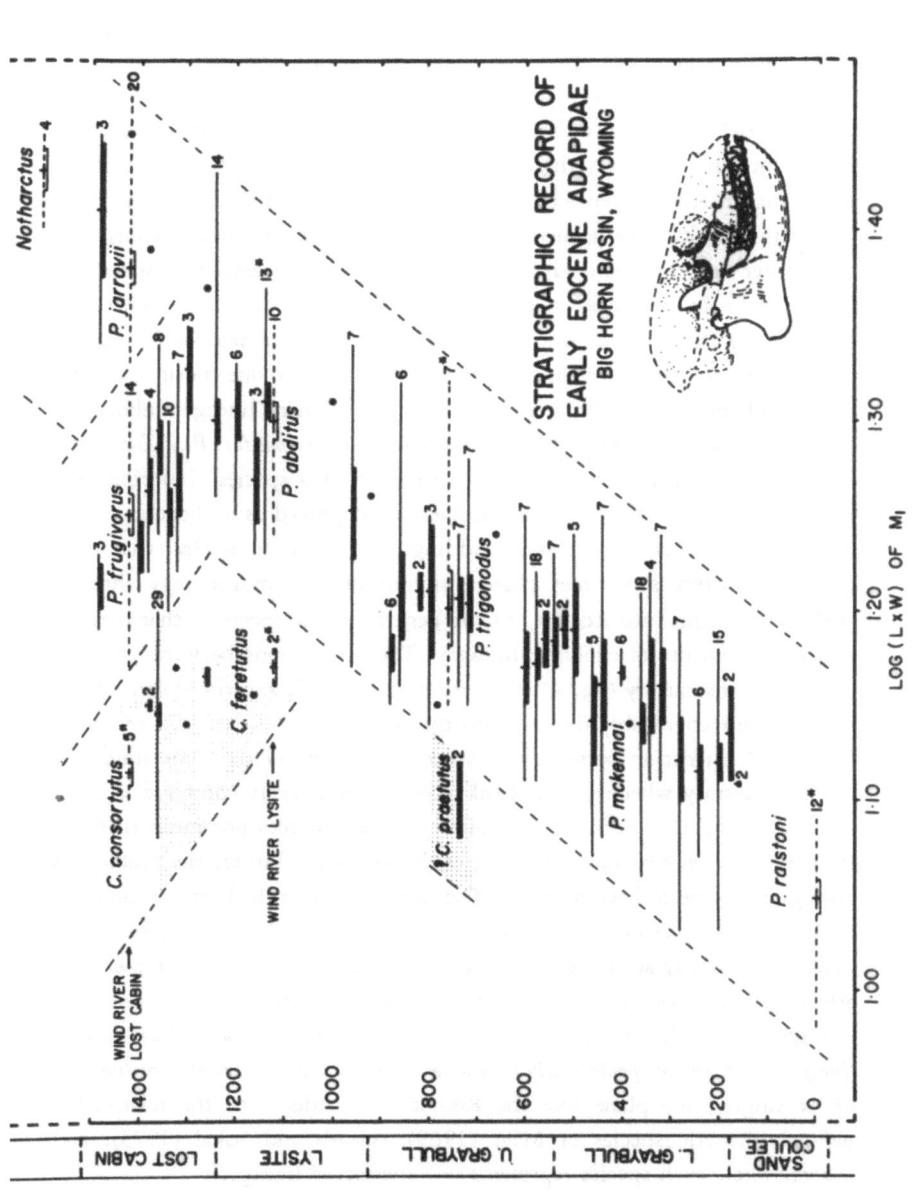

toward larger size from level 200 to level 540. We are looking at change over a period of about 1 to 1.5 million years at one geographic location, hence this is a good example of Simpson's chronocline concept.

Change in *Pelycodus*, as best it can be resolved at present, is both continuous and gradual. Evidence that the change is continuous and gradual is given by the fact that wherever the record is sufficiently dense, there is no statistically significant difference between adjacent samples. Over larger intervals, however, significant differences do accrue as the sum of numerous insignificant differences between adjacent levels. Hence the sample of specimens labeled *Pelycodus trigonodus* in figure 3 is recognizably different, even to the unaided eye in the field, from most samples of *P. mckennai* or *P. abditus*, even though intermediate samples connect them all into one continuous temporal gradient. The result is a continuous evolutionary lineage, subdivided arbitrarily into the segments *Pelycodus mckennai*, *P. trigonodus*, *P. abditus*, etc.

Thus we come to a more general definition of a species. Ignoring geographic and temporal variation, a species is recognized as a phenetic cluster distinct from other such clusters. Adding a time axis, it is clear that these phenetic clusters sometimes change significantly in morphology through time. Taking this into account, a more general idea of species is that they are arbitrary segments of evolving lineages. The distinctions between two lineages are nonarbitrary (fig. 4); their intersection with any time plane shows them to be recognizably distinct clusters, as discussed earlier. On the other hand, the distinction between two successive species of the same lineage is usually arbitrary whenever the fossil record is sufficiently complete to show the transition. This need for an arbitrary boundary does not mean that differences between two successive species do not exist. Rather, the problem is analogous to one of keeping time. The fact that one o'clock grades continuously into two o'clock does not mean that the two hours are the same; we make a necessarily arbitrary boundary between them and recognize that they are different. Combining relational phenetic clustering with temporal change, we can define a species as *an arbitrarily divided segment of an evolving lineage that differs morphologically from other species in the same or different lineages*. On a single time plane like the Recent, this reduces to the relationally defined phenetic species of Mayr (1969), Sneath and Sokal (1973), and others, where each species represents one transected lineage.

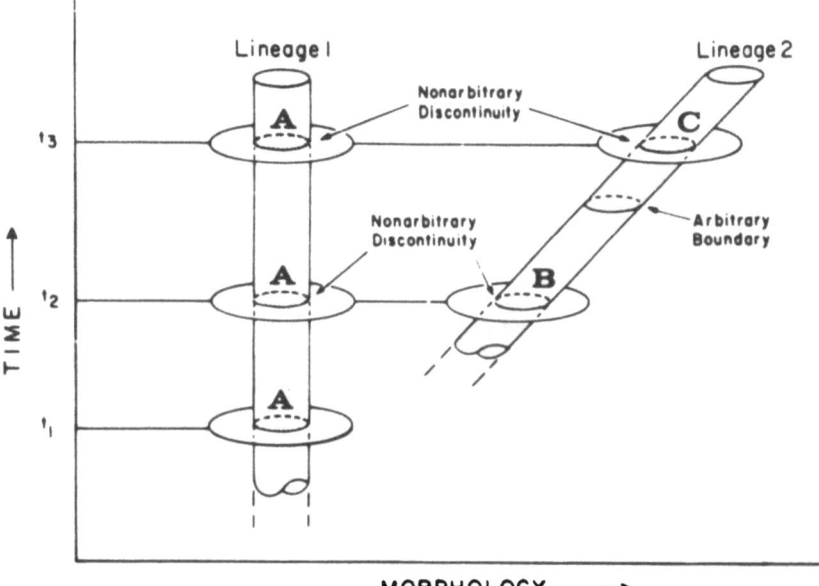

Figure 4. Diagram showing the relationship between biological species (e.g., A and B) sampled at any given time $t$ and generalized paleontological species viewed as units of evolutionary lineages (e.g., B and C). Species in the fossil record have both a biological dimension (where nonarbitrary morphological discontinuities are used to infer reproductive isolation on any given time plane) and a time dimension (where successive morphologically changing units of an evolving lineage must be separated at an arbitrary, time-parallel boundary). In some instances, lineage branching points furnish natural, nonarbitrary boundaries between successive units of evolving lineages. Figure is from Gingerich (1976b).

# Phylogeny Reconstruction

The empirical pattern of morphological distributions through time shown in *Pelycodus* (fig. 3) illustrates an approach to phylogeny reconstruction that uses direct stratigraphic superposition to document the relative temporal position of fossil localities completely independently of the *Pelycodus* in those faunas. This approach I have termed "stratophenetic" because it combines detailed stratigraphic information with phenetic clustering to give an empirical record of phylogeny. The approach requires a relatively dense and continuous fossil record. Where this record is available, a stratophenetic approach gives the most direct and complete reading of phylogeny possible.

## DATA ORGANIZATION

The stratophenetic method can be broken down into three general stages: data organization, phenetic linking, and testing. Data organization can itself be further subdivided. First, each fauna from each fossil locality or the total fauna from each stratigraphic interval must be studied as a distinct unit. Within each sample or stratigraphic interval, specimens are clustered phenetically to determine the total number of species present at that level and which specimens represent the particular species or group of species of interest. This is essentially the same problem as determining the number of species of wild *Canis* present today in the Upper Peninsula of Michigan (fig. 1).

The second aspect of data organization involves arranging all locality samples or faunal intervals in the proper order based on demonstrated superpositional relationships in the field. In some studies, like our present studies in the Sand Coulee area of Wyoming, all fossil samples can be tied directly to measured sections that can be correlated on the basis of lithostratigraphy alone. In other studies, like that of *Pelycodus* in figure 3, many fossil localities were tied directly into a stratigraphic section, but others were interpolated based on stratigraphic and geographic position, and on the morphology of a fossil evolving independently of *Pelycodus* (in that particular case, *Hyopsodus*). Another possibility is to build up a series of stratigraphic sections documenting superpositional relationships of species of interest and then correlate all faunas containing the same phenetic clusters. This approach uses the classic principle of faunal succession and correlation, and it was used most recently in the study of *Plesiadapis* evolution in North America (figs. 5 and 6). A fourth possibility is to break almost entirely biostratigraphical; by postulating a sequence of faunal zones based on the evolution of one group in the fossil record, it is possible to use the resulting zonation to study other independently evolving lineages within this zonal context. For example, the age of

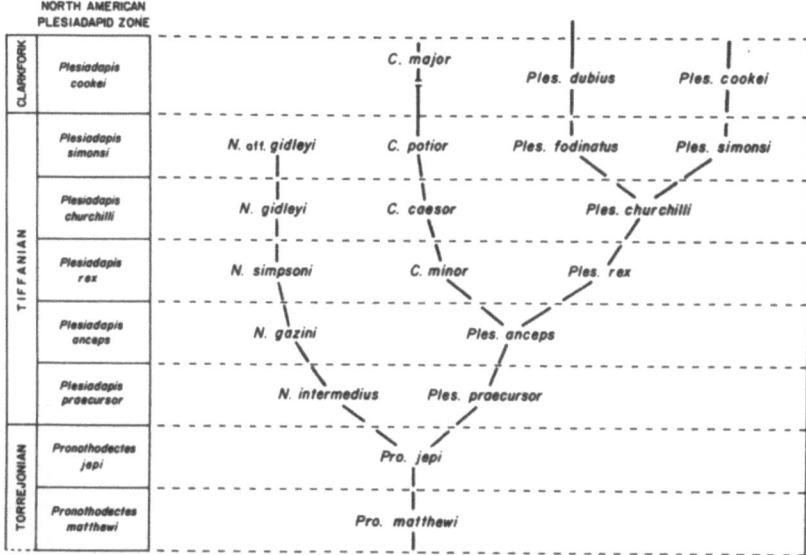

Figure 5. Pattern of stratophenetic linking in North American Paleocene primates of the family Plesiadapidae. Specimens in each locality sample were first clustered phenetically into "biological" species, then the zones were ordered on the basis of stratigraphic superposition, and finally links between species in adjacent zones (heavy lines) were calculated based on overall similarity of all characters available for study. The result is a minimum spanning tree with relative temporal position on the vertical axis. Further discussion and original figure are in Gingerich (1976b).

Eocene mammal faunas in Europe was first inferred from the stage of evolution of perissodactyls in the faunas. Subsequently, the zonation based on perissodactyls was further refined by looking at the evolution of rodents through the same sequence. Most recently, I studied the evolution of Eocene primates in Europe in the context of *niveaux repères* based on the perissodactyls and rodents (see fig. 7).

I have listed four possible methods of ordering locality samples or faunal intervals based on methods that range from purely lithostratigraphic to purely biostratigraphic. Although the former methods are to be preferred, biostratigraphic ordering is also valuable if the ordering is based on taxa evolving independently of those under study (as in *Hyopsodus* relative to *Pelycodus,* or European perissodactyls and rodents relative to primates). If the ordering of samples biostratigraphically is based on taxa other than those under study, the methodology does not involve circular reasoning. However,

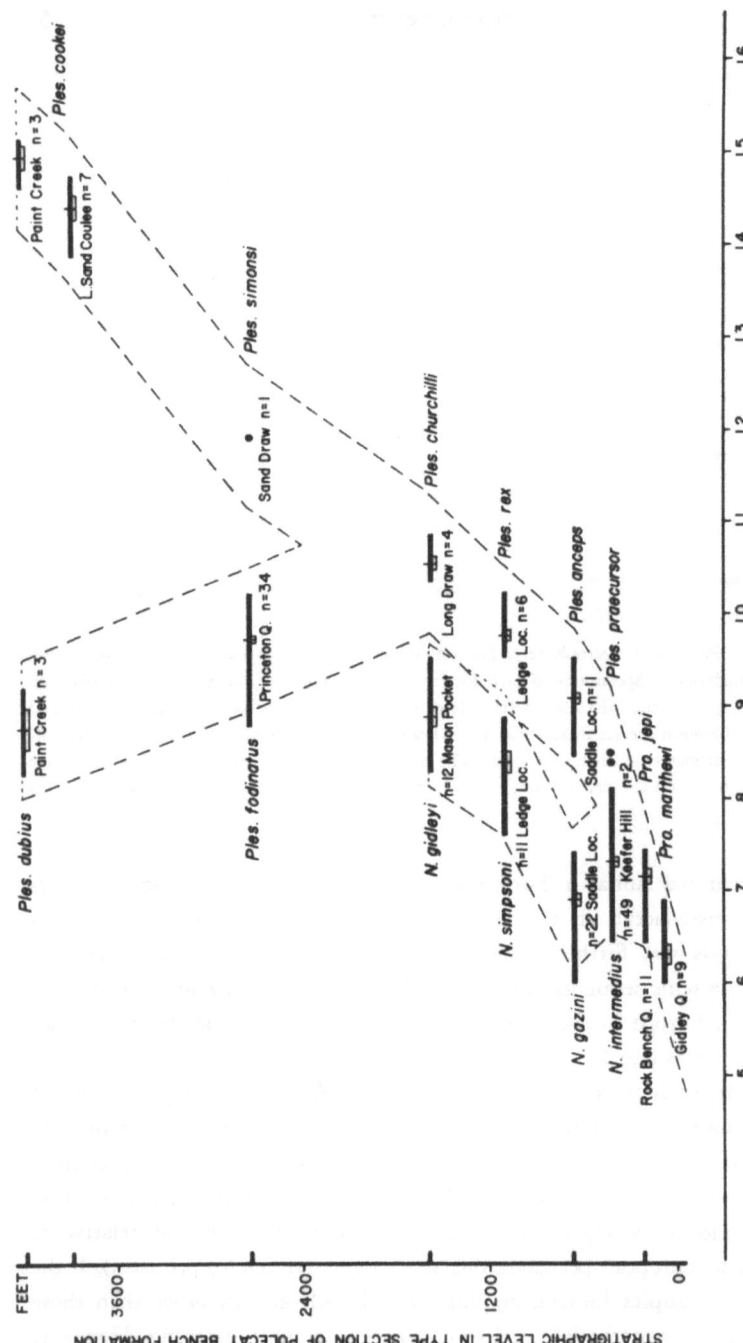

Figure 6. Evolutionary change in one character, relative body size as measured by tooth size, in the radiation of North American species of Plesiadapidae shown in the phylogeny of figure 5. Vertical axis is stratigraphic position; solid bar is sample range; vertical slash is mean; open rectangle is standard error of mean. Figure is from Gingerich (1976b).

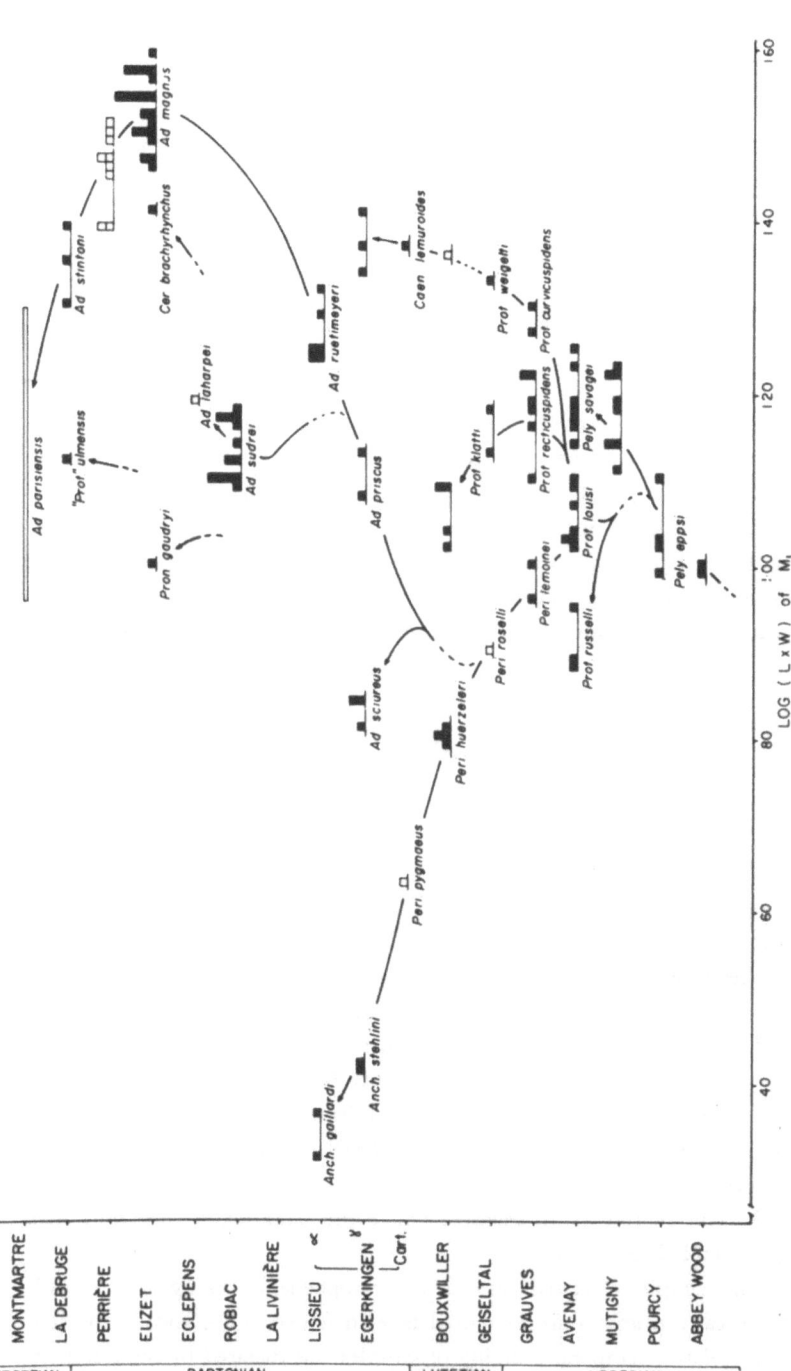

Figure 7. Pattern of stratophenetic linking and evolutionary radiation in European Eocene primates of the family Adapidae. Vertical axis is sequence of *niveaux repères* for the European Eocene, based on perissodactyl and rodent evolution. Horizontal axis is tooth size and, by inference, relative body size. Dashed lines are less well-documented links, showing that some uncertainty persists in this phylogeny owing to gaps in the fossil record. Figure is from Gingerich (1977), with permission of S. Karger AG, Basel.

a purely lithostratigraphic ordering of samples is to be preferred whenever possible, since this obviously involves fewer assumptions about age correlations or evolutionary independence.

## PHENETIC LINKING

Phenetic linking is the second stage of stratophenetic phylogeny reconstruction. Once the phenetic clusters or species in each stratigraphic interval are distinguished and all stratigraphic intervals are arranged in chronological order, then a species in a chosen level can be linked to other species in adjacent levels based on overall similarity. A simplified example has three levels, 1, 2, and 3, superposed one above the other with species A, B, and C in each level (the primes A', A", and B' denote slight morphological differences from typical A and B, respectively):

| Level 3 | A" | B  | C |
|---------|----|----|---|
| Level 2 | A' | B' | C |
| Level 1 | A  | B  | C |

We can first look at A' in Level 2 and ask which species in Level 1 it most closely resembles (answer: A) and then which species in Level 3 it most closely resembles (Answer: A"). Thus A–A'–A" can be linked together on the basis of their overall similarity. Each successive stage represents a progressive change in the original sample A. Similarly B–B'–B can be linked together, with B' interpreted as a character innovation developed in the original B that was subsequently lost again by the time of Level 3. The clearest case of linking comes when C in Level 1 is linked to C in Level 2 and in turn to C in Level 3. This represents an example where the same taxon C appears unchanged in all three levels.

Several examples of actual patterns of stratophenetic linking in the fossil record of early primates are presented here in figures 3, 6, and 7. In each case, the relative temporal ordering of samples was dictated by independent lithostratigraphic or biostratigraphic evidence. These examples differ among themselves in terms of density and continuity of the available record.

More than 500 specimens of early Eocene *Pelycodus* are known from the Big Horn Basin of Wyoming. Some 245 of these preserve $M_1$ and come from known stratigraphic intervals, and these are plotted in figure 3. Each successive species is known from many samples within it which show that successive species grade continuously into each other. In North American Plesiadapidae (fig. 6), all of the species necessary to form a connected phylogeny are known, but here the density of the record is somewhat less, which makes it difficult to study the transitions from one named species to another. The European Adapidae (fig. 7) have a record that is less dense than North American Plesiadapidae but still relatively better known than that of many other groups of mammals. Many lineages, from *Protoadapis louisi* to *Anchomomys gaillardi*, for example, can be traced confidently because all of the species necessary to form a continuous connected phylogeny are known. Other lineages (leading to *Pronycticebus gaudryi* or *Cercamonius brachyrhynchus*, for example) are much less well known.

## HYPOTHESIS TESTING

The final stage of stratophenetic phylogeny reconstruction is hypothesis testing. No matter how well documented a phylogeny may be in terms of density and continuity of the fossil record, there is always the possibility that new fossils will be discovered. Where the fossil record is as dense and continuous as it is for North American *Pelycodus* or even for *Plesiadapis*, it is unlikely that new discoveries will significantly alter the basic phylogenetic pattern proposed. Less well-documented hypotheses, like those indicated by dashed lines in figure 7, are more likely to be modified by future discoveries. Thus it is possible to gain some idea of the relative reliability of various phylogenetic hypotheses by considering the density and continuity of the fossil record on which they are based.

Phylogenies based on a dense and continuous fossil record are usually very stable and reliable. Since stratophenetic hypotheses are constructed using all of the morphological and stratigraphical information available at the time, they cannot normally be tested without the addition of significant new information from the discovery of additional fossil specimens or additional stratigraphic sequences of fossils. This does not mean that a stratophenetic hypothesis is untestable; it is, in fact, tested every time a new fossil belonging to the group in question is found. The fact that new specimens of *Plesiadapis simonsi* morphologically intermediate between *Ples. churchilli* and *Ples. cookei* have been found only in the stratigraphic interval between the latter two species (fig. 6) is additional corroboration of the hypothesis that *Plesiadapis churchilli* gave rise to *Ples. simonsi* which in turn gave rise to *Ples. cookei*.

## ANCESTORS

The sequences of species from *Pelycodus ralstoni* to *Pelycodus jarrovii* in figure 3, from *Pronothodectes matthewi* to *Plesiadapis cookei* in figure 6, and from *Periconodon lemoinei* to *Anchomomys gaillardi* in figure 7 are all examples of ancestor–descendant sequences constructed using stratophenetic methods. An ancestor in this sense, at this level of resolution, is a sample that is temporally, geographically, and morphologically very similar to another species but slightly older in age.

Ancestor–descendant sequences are not always a simple two-species pair. Figures 3, 6, and 7 all illustrate more complicated dichotomies involving three species. *Pelycodus abditus* links very closely to both *Pelycodus frugivorus* and *Pelycodus jarrovii*, and it is probably the ancestor of both. *Plesiadapis churchilli* links very closely to *Plesiadapis fodinatus* and *Plesiadapis simonsi*, and it is probably the ancestor of both. *Protoadapis louisi* links very closely to *Periconodon lemoinei*, *Protoadapis recticuspidens*, and *Protoadapis curvicuspidens*, and it may be the ancestor of all three.

Thus far we have dealt with ancestors in the strict sense, i.e., direct ancestors. A species can be an ancestor in a more general sense. For example, *Pronothodectes matthewi* is ancestral to *Plesiadapis cookei*, and *Pelycodus eppsi* appears to have been ancestral to *Adapis parisiensis*. Thus, at the generic level, we can say that *Pronothodectes* gave rise to *Plesiadapis* and that *Pelycodus* gave rise to *Adapis* in the Old World and *Notharctus* in the New World. These statements can be made with some confidence because most or all of the connecting species are known. In other cases, where connecting sequences of species are not known, it is sometimes possible to make similar general statements about the relationships of genera or even families, although these statements are always less reliable than those regarding species. Obviously, when one says that a family is ancestral to some later, higher taxon, the ancestry is conceived of in a much more general way, with the admission that evidence available at present is insufficient to make a more specific statement about which particular species, or even genus, is actually the ancestor. This problem will be illustrated in discussing the origin of higher primates.

## CHARACTER ANALYSIS

At this point, it is useful to stop and consider in greater detail how the characteristics of species in the fossil record are being analyzed and how phenetic similarity is being measured. Analysis of phenetic similarity is a whole subject in itself (Sneath and Sokal, 1973), and the only aspect I want to consider here is the nature and method of analysis of morphological characters.

When an animal is described, its essential characteristics are enumerated. Each of these characteristics can be used as a taxonomic *character*. Two related species can usually be described using the same set of characters, but the *states* of some individual characters will differ between the two. If all states of all available characters are the same in the two groups and if they are of the same geological age and provenance, the two can rightly be considered one species. Useful characters have at least two states, and, in some cases, three or more states can be defined for a single character. Some examples of the latter can be arranged in a cline ranging from one extreme to the other. To take one simple example, the number of premolars in the mandible of early primates varies, some primates having 2, some 3, and some 4. Premolar number is a character with three states, states 2, 3, and 4, respectively.

Once characters and their states are defined, it is possible to calculate a coefficient of resemblance for two species in a variety of ways. The simple proportion of character states shared by the two would give a crude but direct measure of their similarity. More elaborate measures of distance, association, and correlation for quantifying phenetic resemblance are described by Sneath and Sokal (1973).

The important point about phenetic character analysis is that it does not require any *a priori* assumptions about whether a given character state is primitive or advanced. The direction of an evolutionary change makes no difference in measuring phenetic affinity. Polarity of morphoclines, if they have a simple polarity, is given by relative stratigraphic position.

## Phylogeny at the Species Level

One of the most important reasons for studying phylogeny at the species level is to gain a better understanding of the dynamics of speciation. Paleontology is uniquely suited for study of speciation in natural populations in a natural environment because of the time dimension contributed by the fossil record. There has long been disagreement about the nature of species and consequently about the tempo and mode of speciation. This disagreement has been over whether species are stable and fixed or dynamic and changing. Plato, Cuvier, Goldschmidt, and Schindewolf viewed species as fixed entities, whereas Lamarck, Darwin, and others viewed species as dynamic and changing. As a result, two schools of thought on speciation have grown up, one of which might be called macrogenetic and the other microgenetic. Macrogenesis is the sudden origin of new species or new higher taxa through a saltation of some kind. Microgenesis, on the other hand, refers to a gradual origin of species and new higher taxa by a continuous process of change. Both are possible, and Simpson (1953), for example, argued vigorously against Schindewolf and Goldschmidt while himself introducing the concept of quantum evolution. Most recently, Eldredge and Gould (1972) have again taken up macrogenesis, using discontinuities in the fossil record as their primary evidence (see also Gould and Eldredge, 1977).

Eldredge and Gould's essay on "punctuated equilibria" appears at once to be the most plausible statement of macrogenesis and at the same time open to query. They present "punctuated equilibria" as an alternative to more orthodox phylogenetic gradualism, stating that new species arise rapidly by the splitting of lineages. Eldredge and Gould envision a small isolated subpopulation in a very small peripheral part of the ancestors' geographic range giving rise to each new species. As a consequence, Eldredge and Gould postulate that (1) the first appearance of a new descendant species should be marked by a sharp morphological break between it and its ancestor and (2) breaks in the fossil record reflect discontinuities in evolution (the "punctuation") rather than discontinuities in sedimentation. They further see most normal species populations as being self-regulated by genetic homeostatis and hence essentially unchanging through time (producing the "equilibria"). The hypothesis of punctuated equilibria, as originally stated and illustrated, is an alternative to phyletic gradualism sufficiently explicit to be tested in the fossil record, and I have attempted to do this using the remarkably complete fossil record of early Tertiary mammals preserved in the northern Rocky Mountain area (Gingerich, 1976a,b). One of the examples discussed in that paper can be developed somewhat further.

*Pelycodus* is the most common primate genus found in early Eocene sediments of Wyoming. Originally described from New Mexico by Cope in 1875, the genus was first extensively revised by Matthew (1915), and most recently by Gingerich and Simons (1977) based on new collections made in Wyoming. Matthew's (1915) revision is very interesting and instructive, and the data presented above in figure 3 have been rearranged in figure 8 to illustrate his viewpoint. Matthew divided the main group of early Eocene *Pelycodus* (the lineage from *P. ralstoni* to *P. jarrovii* in fig. 3) into four species: *P. ralstoni*, *P. trigonodus*, *P. abditus* (= *"jarrovii"*), and *P. jarrovii* (= *"Notharctus venticolis"*). Using morphological criteria alone (a practice still advocated by many paleontologists), Matthew diagnosed *P. ralstoni* as having $M_{1-3}$ 11–14 mm in length; *P. abditus*, $M_{1-3}$ 16–18 mm in length; and *P. jarrovii*, $M_{1-3}$ 18–19.6 mm in length. Hypocone enlargement was used as a character also but, since it shows a similar pattern, correlated with overall size, will not be further discussed. Translating Matthew's $M_{1-3}$ lengths into $M_1$ lengths (an equivalent measurement that can be made on many more specimens), the size ranges of the four species in question for $M_1$ were 3.3–4.2, 4.2–4.8, 4.8–5.4, and 5.4–5.9 mm, respectively (fig. 8). The $M_1$ lengths for all specimens ($n = 217$) in these groups in figure 3 have been

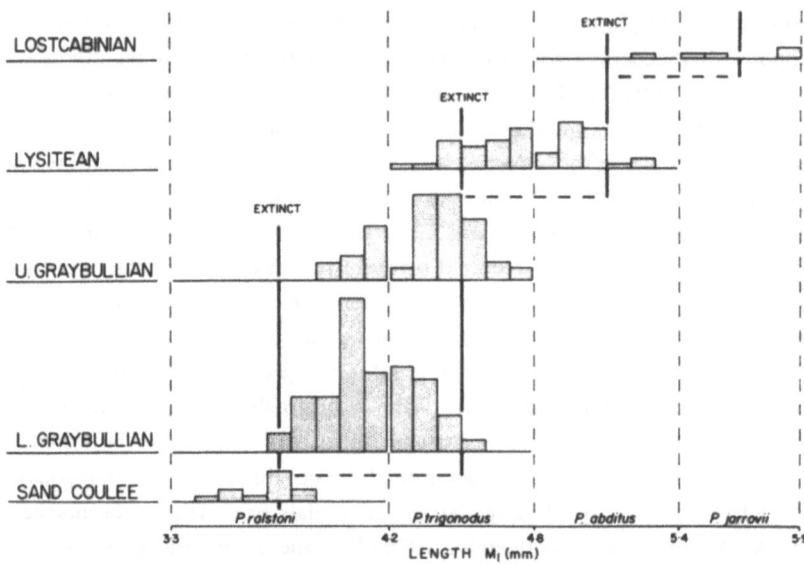

Figure 8. Phylogeny of North American *Pelycodus* as interpreted following the approach of Matthew (1915). Matthew defined species typologically, based only on their morphology, and an erroneous "punctuated" pattern of phylogeny was the result. Finer stratigraphic subdivision and a "biological" approach to the study of variation within each stratigraphic interval yields the single, gradually evolving lineage from *Pelycodus ralstoni* to *Pelycodus jarrovii* shown in figure 3 (which is based on the same data shown here).

replotted by stratigraphic subdivisions of the Wasatch (early Eocene) in figure 8. This figure, showing Matthew's species and stratigraphic subdivisions updated using new data, forms the basis for the following discussion.

According to Matthew (1915), a single species, *Pelycodus ralstoni*, was present in the Sand Coulee beds. By his criteria, it is clear that two species, *P. ralstoni* and the new *P. trigonodus*, were present in both lower and upper Graybull beds, where the former became extinct. *Pelycodus trigonodus* persisted into the Lysitean, where it was joined by the new species *P. abditus*. *Pelycodus trigonodus* became extinct, but *P. abditus* persisted into the Lostcabinian, where it was joined by the new species *P. jarrovii*. *Pelycodus jarrovii* continued on into the middle Eocene, probably giving rise to *Notharctus*. In sum, the evolution of *Pelycodus* in the early Eocene involved the sudden appearance of four species in succession, each of which survived little changed as the sympatric congener of a different, subsequently arriving species. This is, I think, a classic example of punctuated equilibria (see fig. 8), and as such, the example deserves closer study.

One obvious prediction that can be made from the evidence presented in figure 8 is that two sympatric species (a persistent ancestor as well as its descendant and future replacement) are found in each stratigraphic interval. These species, each being essentially static through its known stratigraphic range, should be found side by side in at least some of the better sampled localities. Surprisingly, this is never the case. The range and distribution of variation within a sample from any individual locality containing members of the *P. ralstoni*–*P. jarrovii* lineage is never significantly different from that expected within a single biological population.

Comparing the sequence of samples in the lineage from *P. ralstoni* to *P. jarrovii* in figure 3 with that in figure 8 reveals some interesting differences. Taken level by level, with single specimens omitted, the means of successive samples do not differ significantly from each other at the $p = 0.05$ level except at three places in the entire sequence from *P. ralstoni* to *P. jarrovii*: between levels 880 and 960, levels 960 and 1240, and levels 1240 and 1480. These morphological jumps correspond to breaks in the fossil record, although even here one or two single specimens do bridge the morphological and stratigraphical gap in each case. Variation within any single sample is no larger than that found in single species (largest $V = 6.7$, for the seven specimens from level 960). Thus when the apparently punctuated pattern shown in figure 8 is considered in a more detailed stratigraphic framework, it is clearly a gradual phyletic sequence.

There is another important difference between figures 3 and 8. The species in figure 8 are defined solely on morphological criteria, whereas those in figure 3 are arbitrarily divided at time lines. Even if the stratigraphic record in figure 8 were as finely divided as possible (fig. 3), the evolution of *Pelycodus* would appear, artificially, to be punctuated simply as a consequence of morphological species definitions. Whenever species are diagnosed by arbitrarily dividing the range of variation of a variable character (typology), the result will be an artificially inflated number of sympatric species and a fossil record that appears punctuated. Since selection cannot operate on characters that do not vary, for all practical purposes this problem of typology applies whenever successive species of a lineage are diagnosed morphologically. Hence the patterns of punctuation and equilibrium that appeared in previous studies of *Pelycodus* phylogeny now appear as an inevitable result of purely morphological and typological species definitions and of an insufficiently divided stratigraphic matrix. As it now stands (fig. 3), the phylogeny leading from *Pelycodus ralstoni* to *P. jarrovii* includes three minor "punctuated" saltations, but these coincide with gaps in the fossil record. Thus gaps in the fossil record are a third potential source of artificial punctuation in an otherwise continuous and gradual phyletic sequence.

The sequence from *Pelycodus ralstoni* to *P. jarrovii* forms one continuous, gradually changing lineage with successive species grading perfectly into one another. These species were recognized as being distinctive and worthy of recognition by Matthew and others over 50 years ago, but it is only much more recently that the transition between the species can be documented using much finer stratigraphic detail. There is little question that continuous and gradual evolution in place, *anagenesis,* is important in the origin of some new species. However, punctuated equilibrium is also a hypothesis about the origin of new lineages, and it remains to consider the tempo and mode of origin of new lineages.

Eldredge and Gould (1972) proposed that all new species arise via cladogenesis, the origin of new lineages. Each species itself, in their view, is a separate disconnected lineage. We have seen that this is not always the case, since anagenesis is important in species origins. Now we can consider the geometry of cladogenesis. The *Pelycodus* example in figure 3 shows *P. abditus* giving rise to both *P. jarrovii* and *P. frugivorus.* As discussed above, *P. jarrovii* is a continuation of the *P. ralstoni* to *P. abditus* lineage, with a gap in the fossil record between levels 1240 or 1300 and level 1480. The transition from *P. abditus* to *P. frugivorus,* on the other hand, is much better known.

There is an almost continuous sequence of samples connecting *P. abditus* at level 1240 with the latest *P. frugivorus* samples at level 1480. This forms a perfectly gradual phyletic sequence except for one minor and statistically significant break. There appears to be a significant size difference between the sample of 10 specimens at level 1340 and the sample of 8 specimens directly above it at level 1360 ($t = 2.29$, $p = 0.04$). Interestingly, this shift in means corresponds exactly to the great abundance of the related genus and species *Copelemur consortutus* at level 1360 and may possibly represent an example of character displacement in "ecological" time. After the interval represented by level 1360, *Pelycodus frugivorus* continued to diverge gradually from *P. jarrovii.* This divergence of *P. frugivorus* and *P. jarrovii* also appears to be an example of character displacement, but on a different and "geological" time scale. More study is required to determine if divergences in ecological and geological time are qualitatively different or just different expressions of the same phenomenon.

In conclusion, stratophenetic study of one of the best sequences of fossil mammals available to date shows that the origin of new species of *Pelycodus*, whether via anagenesis or cladogenesis, was a gradual process. Apart from three gaps in the fossil record and a minor possible displacement by a related genus and species, the evolution of *Pelycodus* in North America shows no statistically significant difference from one stratigraphic level to the next, even though the overall trends are highly significant.

# Origin of Major Groups of Vertebrates

The stratophenetic method of phylogeny reconstruction was developed using detailed stratigraphic studies of evolution at the species level, but it is also applicable to studies of evolution at higher levels. Several studies of the evolution of higher groups of vertebrates will be cited from the literature to illustrate the importance of stratigraphic data organization and phenetic linking in reconstructing phylogeny.

## FISH—TETRAPOD TRANSITION

The general aspects of the transition from fish to tetrapod are well understood, in spite of continuing disagreement concerning some details of the transition. Primitive stegocephalian amphibians were highly diversified by the early Carboniferous, some 340 my (million years) before present. Cope and Baur first suggested that early amphibians were derived from Devonian crossopterygians (Schaeffer, 1965a), and discovery of the appropriately named *Ichthyostega* Säve-Söderbergh (1932) in the late Devonian of Greenland provided an important confirmatory link between the two groups. As is usual in well-documented transitions, *Ichthyostega* shows a mosaic of primitive, intermediate, and advanced features linking it phenetically to earlier rhipidistian crossopterygians, on one hand, and to later labyrinthodont amphibians, on the other. A rhipidistian notochordal canal is retained; the intracranial articulation of rhipidistians is absent, but a distinct suture marks its former presence; and postcranially the limbs are well developed. Romer (1964), Schaeffer (1965b), and Thomson (1966) have all considered this transition in more detail. Thomson (1967, p. 672) emphasized the importance of comparing stratigraphically proximate samples by noting that "comparison of finely detailed structures in Recent amphibians and Devonian Rhipidistia is of limited value if information concerning the same structures in the Stegocephalia is not available." In other words, the best evidence concerning phylogenetic relationships of amphibians to other vertebrates comes from comparison of the earliest amphibians with phenetically similar but slightly earlier fishes.

## ORIGIN OF REPTILES

The fossil record makes an essential contribution to many interesting problems of reptilian phylogeny. One of these is the origin of the Class Reptilia as a whole, and here the fossil record has recently provided new and very important evidence. In reviewing problems of reptile origins, Carroll (1970, and fig. 9) has arranged the various groups in question in stratigraphic order. Evidence of phylogenetic relationships can be derived by linking together similar forms of similar age into a coherent pattern. *Romeriscus*, from the early Pennsylvanian (Westphalian A), is one of several important taxa in this regard. According to Baird and Carroll (1967), *Romeriscus* is the earliest reptile yet known, and they state that it could belong to a lineage ancestral to (1) *Seymouria*, (2) romeriid cotylosaurs and pelycosaurs, (3) limnoscelids, and (4) diacodectids. New evidence suggests that *Romeriscus* and the limnoscelids may be anthracosaurian Amphibia rather than primitive Reptilia (fig. 9). The important point is that, looking backward in time, primitive reptiles converge with anthracosaurs in the lower Carboniferous. Forms like *Romeriscus* are difficult to classify precisely because they are transi-

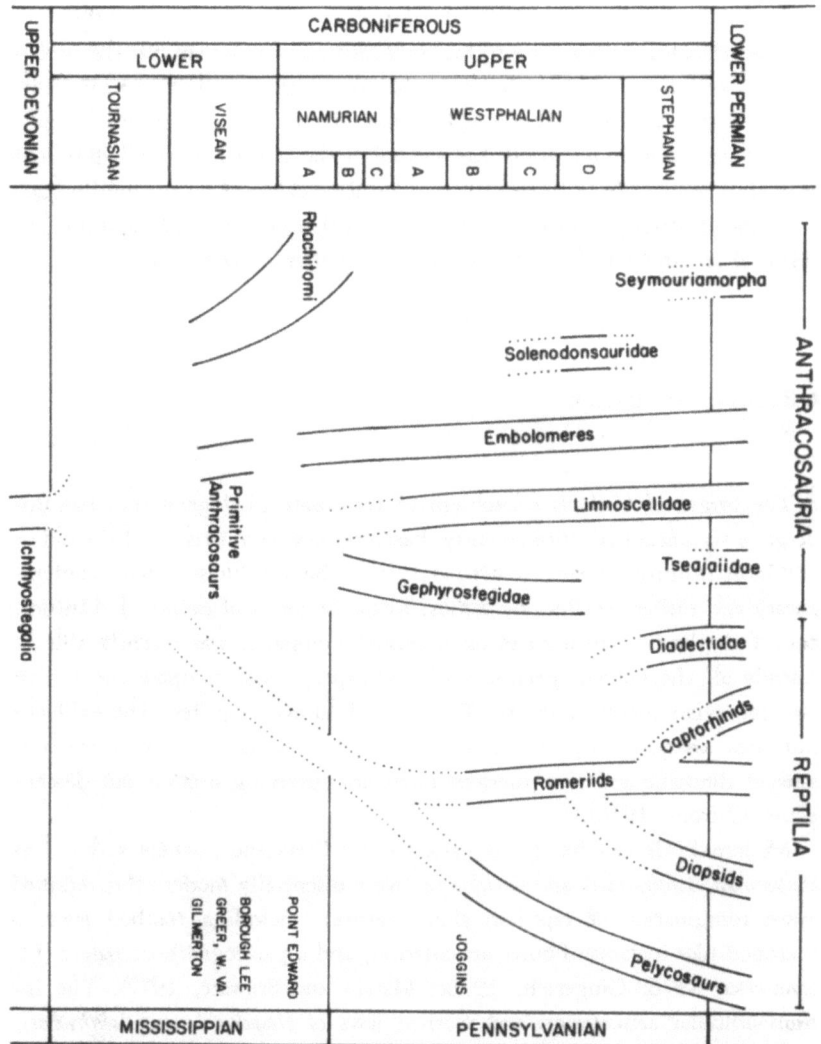

Figure 9. Phylogeny of Paleozoic Reptilia and anthracosaurian Amphibia, showing current knowledge of the amphibian–reptile transition. *Romeriscus*, a primitive limnoscelid from beds of Upper Carboniferous "Westphalian A" age, has been interpreted as either an amphibian or a reptile, and it helps to link amphibian Seymouriamorpha to reptilian Romeriidae and Pelycosauria and to another borderline group Diadectidae. Figure provided by R. L. Carroll (personal communication, 1978).

tional links in a broad evolutionary continuum. Expressing a phylogeny in the traditional manner shown in Figure 9 indicates more than just the probable relationships of critical groups in the amphibian–reptile transition; time ranges of the taxa involved, the approximate time of the transition, and important gaps remaining in our knowledge of the transition are all indicated as well. The fossil record gives the only positive evidence on the morphology and time of origin of primitive reptiles, even though the actual common ancestor of all later reptiles is not known and it may never be known.

## ORIGIN OF BIRDS

The origin of birds is a problem of vertebrate phylogeny that has undergone significant advance recently, based on new comparisons of the oldest and most primitive avian, *Archaeopteryx* from the late Jurassic, with contemporary and earlier reptiles. By a most fortunate series of events, J. Ostrom, one of the leading authorities on theropod dinosaurs, was recently able to restudy all the known specimens of *Archaeopteryx* and compare them with possible avian ancestors among Triassic and Jurassic reptiles. The evidence indicates unequivocally that *Archaeopteryx* evolved from a small coelurosaurian dinosaur and that modern birds are surviving dinosaurian descendants (Ostrom, 1976).

Modern birds can be traced back to the Cretaceous, where some, like *Hesperornis, Ichthyornis,* and *Gobipteryx,* were essentially modern but retained some combination of reptilian skull features, including toothed jaws, a theropod-like intramandibular articulation, and a palaeognathous palate (Elzanowski, 1976; Gingerich, 1976c; Martin and Stewart, 1977). The intramandibular articulation and toothed jaws of *Hesperornis* and *Ichthyornis,* making them morphologically as well as temporally intermediate between theropods and modern birds, are illustrated in figure 10. *Archaeopteryx* too may have had such an intramandibular articulation (although Wellnhofer, 1974, did not find evidence for this), and it certainly had toothed jaws.

*Archaeopteryx* combines a coelurosaurian skeleton with avian feathers in a temporally intermediate morphological mosaic linking birds to a theropod ancestry. As Ostrom (1976, p. 93) has stated:

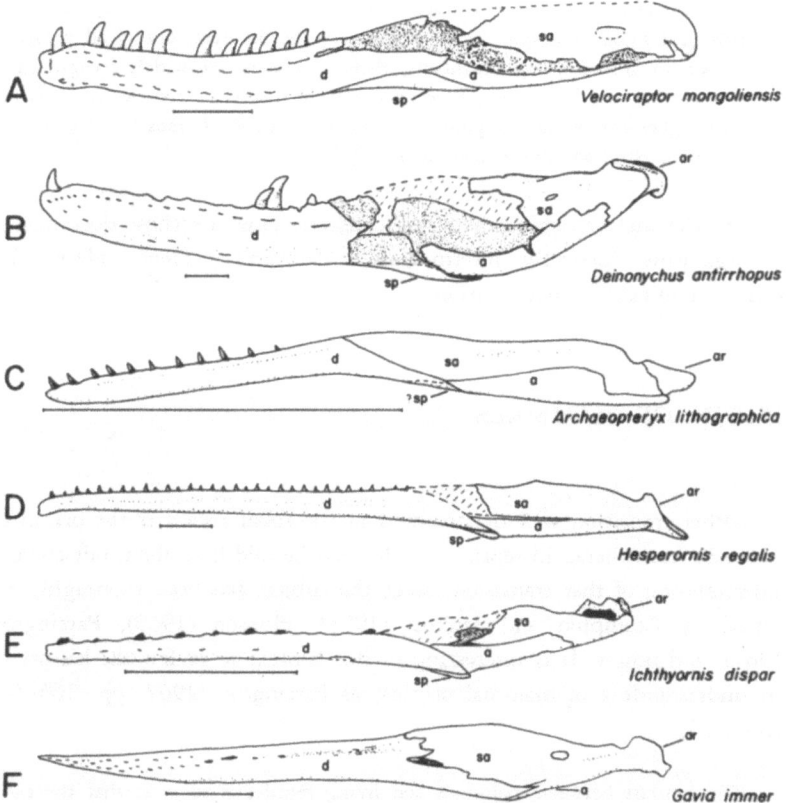

Figure 10. Left mandibles of theropod dinosaurs (A and B), Mesozoic toothed birds (C–E), and a modern bird, the loon (F), all brought to about the same length (scale bar = 2 cm). Mesozoic toothed birds show a mosaic of features intermediate between theropod dinosaurs and modern birds, linking birds to a theropod ancestry. Note teeth, and splenial-angular intramandibular articulation shared by theropods, *Hesperornis*, *Ichthyornis*, and possibly *Archaeopteryx*. *Archaeopteryx* had avian feathers, and both *Hesperornis* and *Ichthyornis* had an essentially modern avian postcranial skeleton. *Velociraptor* is from Osborn (1924) and *Archaeopteryx* from Wellnhofer (1974); other drawings are based on original specimens in collections of the Yale Peabody Museum.

Possibly no other zoological specimens, fossil or Recent, are considered so important as are those of *Archaeopteryx lithographica*. Certainly few other specimens have generated such widespread interest or provoked as much speculation and controversy. The reasons are several: these specimens are the oldest known fossil bird remains; they are extremely rare, only five specimens (excluding the solitary feather) are known at present; several of these preserve remarkably detailed impressions of feathers and an extraordinary mixture of reptilian and avian characters; and most important of all, because of the last fact, out of all presently known fossil and living organisms, these specimens are widely recognized as constituting the best example of an organism perfectly intermediate between two higher taxonomic categories—representing an ideal transitional stage between ancestral and descendant stocks.

Since the most reliable phylogenetic relationships are those documented by organisms "perfectly intermediate," Ostrom's evidence places *Archaeopteryx* in exactly that position.

## ORIGIN OF MAMMALS

Another transition well documented in the fossil record is the origin of mammals from therapsid reptiles. Little need be said here about our current understanding of that transition, since the subject has been thoroughly reviewed by Crompton and Jenkins (1973), Hopson (1969), Parrington (1967), and others. It is interesting in this context to review the history of our understanding of mammal origins, as Parrington (1967, pp. 165–66) recounts it:

> The contrast between mammals and living reptiles is so great that the early zoologists felt that they could not be related and that the origin of the mammals must be sought among the amphibia and not the reptiles. It was found, for example, that the mammalian circulatory system, in which the left systemic arch carries arterial blood from the left ventricle of the heart, could not have been evolved from that of any modern reptile because in these forms it is the right systemic arch which carries the blood of the left ventricle. But both mammalian and reptilian circulatory systems could reasonably be supposed to have evolved from a symmetrical arterial system such as is found in some amphibia. Again the study of the subdivisions of the coelom indicates that the mammalian diaphragm can be derived from the amphibian structures better than those of the reptiles. And such trivial resemblances as the paired exoccipital condyles, found in both amphibia and mammals, were quoted in support of this view.
> When, in 1858, the British Association first met in the city of Leeds it did so under the presidency of that great anatomist Sir Richard Owen who, at the time, was studying the earliest discovered mammal-like reptiles. These had been sent to him from South Africa, and in the following year he described the first

theriodont, a cynodont he called *Galesaurus planiceps*. . . . And gradually an incontestable case was made out to show the evolution of the mammals, step by step, from the primitive Pelycosaurs of the Lower Permian *via* the almost mammalian theriodonts of the Upper Trias.

Clearly, documentation that mammals evolved from early reptiles and not from amphibians came from discoveries of Permo-Triassic "mammal-like reptiles" intermediate in time and morphology. More detailed evidence remains to be discovered, but it is unlikely that the broad link between early mammals and reptiles will be altered significantly.

*Ichthyostega, Romeriscus, Archaeopteryx*, etc., may not be the actual ancestors of the groups they represent, but taking a slightly broader view, it is unlikely that the actual ancestors differed greatly from these genera. Each provides a continuity in time and morphology that links early members of major radiations together.

# Phylogeny of Hominidae

The stratophenetic methodology outlined at the beginning of this paper has been applied, consciously or unconsciously, to problems of vertebrate phylogeny by many students of the subject. These problems range in scale from the origin and relationships of individual vertebrate species to the origin of the entire subphylum Vertebrata. Some examples illustrating the importance of stratigraphically and morphologically similar series of fossils have been given earlier. It remains to discuss the sequence of fossils available that documents our own position in vertebrate phylogeny. A discussion of the general phylogeny of primates using a stratophenetic approach has appeared elsewhere (Gingerich and Schoeninger, 1977). Here I will concentrate on the evidence bearing on hominid phylogeny.

New discoveries are made each year, especially in East Africa, that significantly augment the stratigraphic and morphological evidence bearing on human phylogeny. Until relatively recently, it was possible to argue forcefully that only one lineage of Hominidae existed, a lineage that led from little-specialized Miocene hominoids or hominids to *Australopithecus*, to *Homo erectus*, and finally to modern *Homo sapiens*. By 1964, however, discoveries in the stratified sediments of Olduvai Gorge in northern Tanzania led L. Leakey, Tobias, and Napier (1964, p. 7) to state that

> examination of these finds has enabled us to broaden the basis of our diagnosis of the proposed new species and has fully confirmed the presence of the genus *Homo* in the lower part of the Olduvai geological sequence, earlier than, contemporary with, as well as later than, the *Zinjanthropus* skull, which is certainly an australopithecine.

The past dozen years or so have seen an active debate over single versus multiple lineage hypotheses in human evolution. This basic question was debatable largely because the evidence bearing on it was meager, but fortunately, in recent years much new evidence has been collected from a stratified sequence of sediments in northern Kenya, in the "East Rudolf" area, east of what is now called Lake Turkana. There the evidence is beginning to be complete enough to apply a stratophenetic approach to human phylogeny (see Gingerich and Schoeninger, 1977, for references and discussion). This approach is illustrated in figure 11.

Three stratigraphic intervals are defined in the area east of Lake Turkana. Each is bounded above by a radiometrically dated tuff complex (fig .11). The sub-KBS interval is overlain by a tuff complex dated at about 1.8 my. Above this, the sub-Okote interval is overlain by a tuff dated at about 1.5 to 1.6 my. The sub-Karari interval overlies the Okote tuff and is in turn overlain by the Karari tuff dated at about 1.3 my. Fossil hominids have been described from these three intervals in papers by R. Leakey (1977), R. Leakey and Walker (1977), and others (Day et al., 1975, 1976; papers in Coppens et al., 1976). In addition, M. Leakey et al. (1976) and White (1977) have described a collection of hominids from one locality, Laetolil, that predates the sub-KBS interval east of Lake Turkana, and these specimens are included in figure 11 to give an earlier perspective on hominid phylogeny.

The pattern of change in tooth size illustrated in figure 11 can be analyzed level by level stratigraphically, just as we analyzed patterns in *Pelycodus*, *Plesiadapis*, etc. The hominid sample from the Laetolil interval appears to represent a single biological species, as illustrated by the clustering of tooth size and the great similarity of all other features of the known dentition (cranial and postcranial elements are as yet unknown for this sample). The sub-KBS interval east of Lake Turkana contains specimens that show some tendency to cluster into larger, robust forms and smaller, gracile forms. It is not yet certain that more than one biological species is being sampled in the

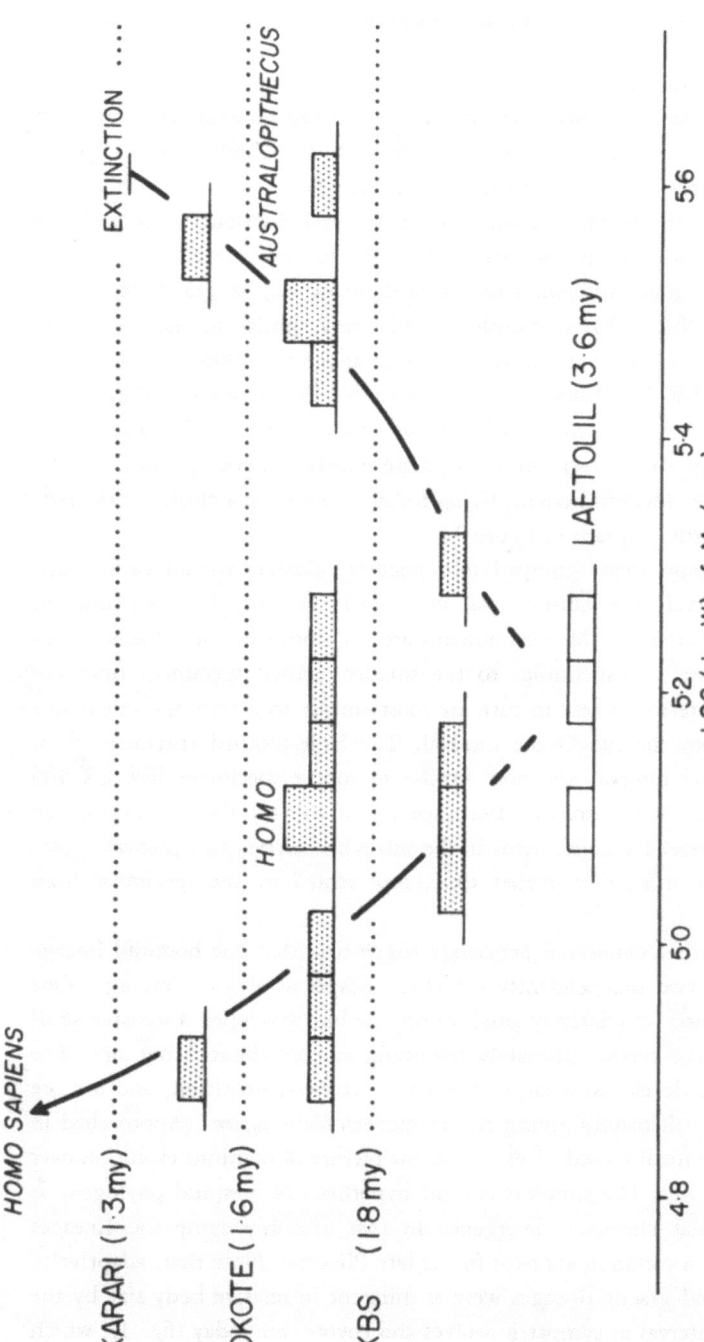

Figure 11. Phylogeny of Plio-Pleistocene Hominidae in East Africa. Upper three intervals are from stratigraphic sections east of Lake Turkana in Kenya; bottom interval is from Laetolil in northern Tanzania (radiometric ages in parentheses). Note marked bimodality in upper levels, with robust *Australopithecus* becoming extinct at about 1 my ago and the gracile *Homo* lineage giving rise to modern *Homo sapiens*. Compare distribution in sub-Okote interval with Canidae in figure 1. See text for discussion.

sub-KBS interval, but the distribution of tooth size and other cranial characteristics suggests that this is so. By the sub-Okote interval, there are very clearly a robust, large-toothed *Australopithecus* (or *"Zinjanthropus"*) species and a more gracile, small-toothed *Homo* species. Differences in tooth size are clear, with both samples showing the range and distribution of variation characteristic of modern mammal species. Fortunately, recently discovered crania confirm these differences in a most dramatic way. A gracile skull from this interval (ER-3733) is morphologically very similar to that of Peking *Homo erectus*, with an endocranial volume of about 800–900 cm$^3$, whereas a robust skull (ER-406) from the same interval has the brain size and morphology typical of *Australopithecus* (R. Leakey and Walker, 1977). Two specimens from the sub-Karari interval separate widely, one being much smaller than the other. Modern humans living today would form a cluster at the left-hand side of the diagram in figure 11.

With the specimens grouped into phenetic clusters within each stratigraphic interval, the clusters can be linked vertically between adjacent stratigraphic intervals. Modern humans are, in tooth size and all other comparable features, most similar to the smaller-toothed specimens from the sub-Karari interval. These in turn are most similar to gracile specimens like ER-3733 from the sub-Okote interval. The large-toothed specimens from the sub-Karari interval are most similar to robust specimens like ER-406 from the sub-Okote interval. Both species in the sub-Okote interval can similarly be traced via specimens in the sub-KBS interval to a probable common ancestor at an even earlier time, represented by the specimens from Laetolil.

The result is a connected genealogy suggesting that one hominid lineage divided into two independently evolving lineages at about 2 my ago. One lineage retained its relatively small brain size but developed a massive skull with very large teeth, ultimately becoming extinct about 1 my ago. The other lineage developed a larger brain size, reduced dentition, and a more gracile skull, ultimately giving rise to modern *Homo sapiens*. Approached in this way, the fossil record gives a coherent picture of hominid evolution over the past 3.6 my. The simplest current hypothesis of hominid phylogeny is one of gradual character divergence in two probably sympatric lineages derived from a common ancestor in the late Pliocene. Note that, adaptively, the robust and gracile lineages were as different in relative body size by the sub-Okote interval as sympatric wolves and coyotes are today (fig. 1), which indicates that body size was probably an important adaptive difference between the two hominid lineages.

It is possible that human phylogeny is still more complicated, since there is some evidence from the size of the brain that perhaps the gracile lineage really included two separate lineages, one with large brains and the other with a brain no larger than typical *Australopithecus* (R. Leakey, 1976). This question could best be studied, when sufficient information is available, by a similar study including brain size on an additional axis in the stratophenetic plot.

# Summary and Conclusions

In the preceding pages I have outlined a paleontological approach to phylogeny reconstruction. This "stratophenetic" method combines stratigraphic evidence of relative temporal position with phenetic clustering and linking to yield an essentially empirical reading of phylogeny. Where the fossil record is dense and continuous, a relatively clear pattern of genealogy emerges, but where there are large gaps in the record, the pattern is often ambiguous, a warning against any strong statement of relationships in such a case. The latter attribute of the stratophenetic method is in fact a positive feature, since it means that an arbitrary phylogeny will not be constructed automatically by a mechanical algorithm unless there is some actual historical evidence on which to base the phylogeny.

The stratophenetic method can be applied to problems of phylogeny at any level, from dealing with relationships of individual populations to comparing classes of phyla. Some examples of phylogeny reconstruction at the species, the family, and class level have been given. In the species-level examples, both *Pelycodus* and East African Hominidae exhibit a pattern of gradual phyletic evolution, with branching events being followed in each case by character divergence in body size. As was shown in *Pelycodus* (fig. 8), this gradual pattern will appear to be "punctuated" if typological species definitions ignoring morphological variability are employed or if stratigraphic intervals are too broadly defined. Gaps in the fossil record will likewise produce an artificial appearance of punctuation.

In considering phylogenetic relationships at a higher level, it was seen that *Ichthyostega, Romeriscus,* and *Archaeopteryx* link classes of Vertebrata together in a general way, although to claim that any one known genus was *the* ancestor of a given class exceeds the level of resolution of the available information. Nevertheless, morphologically and temporally intermediate forms like *Archaeopteryx* give the best evidence on the general relationships of birds to reptiles, and *Archaeopteryx* is unlikely to differ greatly from the actual ancestor of birds.

Finally, it should be emphasized that the stratophenetic approach to phylogeny reconstruction is basically an empirical approach. This approach requires a relatively dense and continuous fossil record, but observation more than philosophy determines the resulting pattern of phylogeny. Dampier (1966, p. xv) has characterized natural science as follows:

> Natural science may use deductive reasoning at an intermediate stage of its enquiries, and inductive theories are an essential part of its procedure, but primarily it is empirical, and its ultimate appeal is to observation and experiment; it does not, like mediaeval Scholasticism, accept a philosophic system on authority and then argue from the system what the facts ought to be.

This is basically the approach taken here. The fossil record of vertebrate evolution is better than many people realize, and it is of fundamental importance for understanding phylogeny. If paleontology were synonymous with morphology, fossils would be far less valuable than they are today. However, paleontology is a historical science, and the stratigraphical attributes of fossils give them a unique importance in the study of phylogeny.

# Acknowledgments

The methodology of phylogeny reconstruction outlined here was developed in the course of detailed stratigraphic studies of Eocene mammals in Wyoming, research supported in recent years by several faculty research grants from the Rackham School of Graduate Studies. Professors J. H. Ostrom and E. L. Simons of the Peabody Museum, Yale University, permitted generous access to collections of fossil mammals, birds, and reptiles figured here. Professor M. H. Wolpoff and Dr. T. D. White, University of Michigan Department of Anthropology, provided much of the data used to construct figure 11. Dr. P. Myers, University of Michigan Museum of Zoology, permitted me to study collections of Sciuridae and Canidae used to construct figures 1 and 2. Many other colleagues as well have helped in developing the ideas outlined in this paper. Mrs. Gladys Newton typed the final manuscript.

# References

Baird, D., and Carroll, R. L. 1967. *Romeriscus*, the oldest known reptile. *Science* 157:56–59.
Carroll, R. L. 1970. The ancestry of reptiles. *Phil. Trans. Roy. Soc. London Ser. B* 257:267–308.
Coppens. Y., Howell, F. C., Isaac, G. L., and Leakey, R. E. F. (eds.). 1976. *Earliest Man and Environments in the Lake Rudolf Basin: Stratigraphy, Paleoecology, and Evolution*. University of Chicago Press, Chicago.
Crompton, A. W., and Jenkins, F. A. 1973. Mammals from reptiles: A review of mammalian origins. *Ann. Rev. Earth Planet. Sci.* 1:131–54.
Dampier, W. C. 1966 *A History of Science*, reprint of 4th ed. Cambridge University Press, Cambridge.
Day, M. H., Leakey, R. E. F., Walker, A. C., and Wood, B. A. 1975. New hominids from East Rudolf, Kenya, I. *Amer. J. Phys. Anthropol.* 42:461–76.
Day, M. H., Leakey, R. E. F., Walker, A. C., and Wood, B. A. 1976. New hominids from East Turkana, Kenya. *Amer. J. Phys. Anthropol.* 45:369–436.
Eldredge, N., and Gould, S. J. 1972. Punctuated equilibria: An alternative to phyletic gradualism. In T. J. M. Schopf (ed.), *Models in Paleobiology*, Freeman, Cooper and Co., San Francisco.
Elzanowski, A. 1976. Palaeognathous bird from the Cretaceous of central Asia. *Nature* (London) 264:51–53.
Gingerich, P. D. 1976a. Paleontology and phylogeny: Patterns of evolution at the species level in early Tertiary mammals. *Amer. J. Sci.* 276:1–28.
Gingerich, P. D. 1976b. Cranial anatomy and evolution of early Tertiary Plesiadapidae. *Univ. Mich. Pap. Paleontol.* 15:1–140.
Gingerich, P. D. 1976c. Evolutionary significance of the Mesozoic toothed birds. *Smiths. Contrib. Paleontol.* 27:23–33.
Gingerich, P. D. 1977. New species of Eocene primates and the phylogeny of European Adapidae. *Folia Primatol.* 28:60–80.
Gingerich, P. D., and Schoeninger, M. J. 1978. The fossil record and primate phylogeny. *J. Human Evol.* 6:483–505.

Gingerich, P. D., and Simons, E. L. 1977. Systematics, phylogeny, and evolution of early Eocene Adapidae (Mammalia, Primates) in North America. *Contrib. Univ. Michigan Mus. Paleontol.* 24:245–79.

Gould, S. J. 1975. On the scaling of tooth size in mammals. *Amer. Zool.* 15:351–62.

Gould, S. J., and Eldredge, N. 1977. Punctuated equilibria: The tempo and mode of evolution reconsidered. *Paleobiology* 3:115–51.

Hopson, J. A. 1969. The origin and adaptive radiation of mammal-like reptiles and nontherian mammals. *Ann. N.Y. Acad. Sci.* 167:199–216.

Hutchinson, G. E., and MacArthur, R. H. 1959. A theoretical ecological model of size distributions among species of animals. *Amer. Nat.* 93:117–25.

Hutton, J. 1788. Theory of the earth, or an investigation of the laws discernible in the composition, dissolution and restoration of land upon the globe. *Trans. Roy. Soc. Edinburgh* 1:209–304.

Leakey, L. S. B., Tobias, P. V., and Napier, J. R. 1964. A new species of the genus *Homo* from Olduvai Gorge. *Nature (London)* 202:7–9.

Leakey, M. D., Hay, R. L., Curtis, G. H., Drake, R. E., Jackes, M. K., and White, T. D. 1976. Fossil hominids from the Laetolil beds. *Nature (London)* 262:460–66.

Leakey, R. E. F. 1977. New hominid fossils from the Koobi Fora formation in northern Kenya. *Nature (London)* 261:574–76.

Leakey, R. E. F., and Walker, A. C. 1977. *Australopithecus, Homo erectus* and the single species hypothesis. *Nature (London)* 261:572–74.

Martin, L. D., and Stewart, J. D. 1977. Teeth in *Ichthyornis* (Class: Aves). *Science* 195:1331–32.

Matthew, W. D. 1915. A revision of the lower Eocene Wasatch and Wind River faunas. Entelonychia, Primates, Insectivora (part). *Bull Amer. Mus. Nat. Hist.* 34:429–83.

Mayr, E. 1969. The biological meaning of species. *Biol. J. Linn. Soc.* 1:311–20.

Mayr, E. 1970. *Populations, Species, and Evolution.* Belknap/Harvard University Press, Cambridge, Mass.

Mayr, E. 1974. Cladistic analysis or cladistic classification? *Z. Zool. Syst. Evolut. Forsch.* 12:94–128.

McNab, B. K. 1971. On the ecological significance of Bergmann's rule. *Ecology* 52:845–54.

Osborn, H. F. 1924. Three new theropoda, Protoceratops zone, central Mongolia. *Amer. Mus. Novit.* 144:1–12.

Ostrom, J. H. 1976. *Archaeopteryx* and the origin of birds. *Biol. J. Linn. Soc.* 8:91–182.

Parrington, F. R. 1967. The origins of mammals. *Adv. Sci.* 1967:165–73.

Romer, A. S. 1964. Problems in early amphibian history. *J. Ani. Morph. Physiol.* 11:1–20.

Rosenzweig, M. L. 1966. Community structure in sympatric Carnivora. *J. Mammal.* 47:602–12.

Säve-Söderbergh, G. 1932. Preliminary note on Devonian stegocephalians from East Greenland. *Medd. Groenl.* 94:1–107.
Schaeffer, B. 1965a. The evolution of concepts related to the origin of the amphibia. *Syst. Zool.* 14:115–18.
Schaeffer, B. 1965b. The rhipidistian–amphibian transition. *Amer. Zool.* 5:267–76.
Simpson, G. G. 1943. Criteria for genera, species, and subspecies in zoology and paleozoology. *Ann. N.Y. Acad. Sci.* 44:145–78.
Simpson, G. G. 1953. *The Major Features of Evolution.* Columbia University Press, New York.
Simpson, G. G. 1975. Recent advances in methods of phylogenetic inference. In W. P. Luckett and F. S. Szalay (eds.), *Phylogeny of the Primates,* pp. 3–19. Plenum Press, New York.
Simpson, G. G. 1976. The compleat palaeontologist? *Ann. Rev. Earth Planet. Sci.* 4:1–13.
Sneath, P. H. A., and Sokal, R. R. 1973. *Numerical Taxonomy.* Freeman and Co., San Francisco.
Sokal, R. R., and Crovello, T. J. 1970. The biological species concept: A critical evaluation. *Amer. Nat.* 104:127–53.
Stanley, S. M. 1973. An explanation for Cope's rule. *Evolution* 27:1–26.
Thomson, K. S. 1966. The evolution of the tetrapod middle ear in the rhipidistian–amphibian transition. *Amer. Zool.* 6:379–97.
Thomson, K. S. 1967. Notes on the relationships of the rhipidistian fishes and the ancestry of the tetrapods. *J. Paleontol.* 41:660–74.
Van Valen, L. 1973. Body size and numbers of plants and animals. *Evolution* 27:27–35.
Wellnhofer, P. 1974. Das fünfte skelettexemplar von *Archaeopteryx. Palaeontographica Ser. A* 147:169–216.
White, T. D. 1977. New fossil hominids from Laetolil, Tanzania. *Amer. J. Phys. Anthropol.* 46:197–230.

# An Introduction to the Logic of Phylogeny Reconstruction

**Eugene S. Gaffney**

> *The game of science is, in principle, without end. He who decides one day that scientific statements do not call for any further test and that they can be regarded as finally verified, retires from the game.*
>
> Popper, *The Logic of Scientific Discovery*

A student being introduced to systematic zoology finds that there are a few standard textbooks by widely esteemed individuals (e.g., Simpson and Mayr), and these are immediately read with great enthusiasm. They seem to explain what one sees systematists doing and claim to derive their success from the synthetic theory of evolution. However, sooner or later, depending on the vitality of the academic environment, the student realizes that a growing proportion of practicing animal systematists do not regard many of the ideas propounded by those authors very highly. Instead he or she is told to recant Simpson and Mayr and count himself with Sokal and Sneath or to recant Simpson and Mayr and split himself off as a disciple of Hennig. Generally speaking, most students may weigh each argument objectively but, in the end, seem to adopt the methodology of the people surrounding them.

Although it may be that the course of systematics is more a problem in the sociology of science than the logic of science, I earnestly hope that this is not so. It seems to me that the most important disagreement among systematists concerns the nature of science and that a resolution of this problem is of primary importance. This insistence on reexamining the nature of science, a condition that is very prominent in systematics at present, has been considered by Kuhn (1970) to mark the first stages of what he calls scientific revolutions (see also Hull, 1975; Nelson, 1975). It may very well be that we are entering a new period in systematics, but in any case, in order to have meaningful discussion of issues, there must be some common ground, and the purpose of my appeal to philosophy is to establish some areas of general agreement.

I hope to provide in this paper a brief introduction to the ideas of phylogenetic systematics (or cladism[1]), as I see them, by developing a scientific methodology and relating phylogeny reconstruction to it. I make no claim of unanimity with other cladists, but on the other hand, few of the points expressed here are original. Nonetheless, this is not a literature review of cladism (for this, see Bonde, 1977), nor is it an attempt to show the diversity of ideas that could come under the subject of cladism.

## The Hypothetico-Deductive Method

More likely than not, if a scientist is pressed about methodology or a graduate student is asked to define or discuss the scientific method, some description of the so-called Baconian view of science will emerge. In this view, the fact-gathering process reigns supreme and is followed by the more or less automatic emergence of an explanatory theory or "law" when the amount of facts or observations reaches a critical level. At this point, observations or experiments are made that attempt to verify or prove the truth of

---

[1] Mayr and others critical of Hennig applied the term "cladism" to this "taxonomic theory" which "depends on the position of branching points on the phylogenetic tree" (Mayr, 1969, p. 70). The characterization of the terms "cladism" and "cladistics" in this somewhat misleading fashion has led many "cladists," including Brundin, Rosen, Nelson, and Cracraft, to ignore these terms in favor of the unusually apt term "phylogenetic systematics," as used by Hennig. There is good precedent for using "cladogenesis" as branching or diversifying evolution (Huxley, 1958; Rensch, 1959); nonetheless, as Huxley (1958, p. 27) has emphasized, the term "clade" refers to "monophyletic units of whatever magnitude." In this sense, I take cladism to mean the study of monophyletic groups, and that is precisely what I am talking about in this paper.

the theory. If the observations meet expectation, then the theory or law is added to the general fund of knowledge believed to be true, and the gathering of observations begins on a new frontier of science. The method of basing theories or general statements on many observations of specific instances is usually called induction and is often hailed as the scientific method.

Some scientists, as well as philosophers, have argued that this method is logically unsatisfactory and, in any case, is not the way scientists work. Hume was one of the most influential philosophers to develop a criticism of induction, and among scientists, it was Einstein who emphasized that "theory cannot be fabricated out of the results of observation" (Magee, 1973, p. 25).

What, then, can we use as a logically consistent methodology, and what methodologies have been used by successful scientists in the past? The simple, yet compelling, answer to this question is: Develop a hypothesis and test it, preferably by trying to show that it is wrong rather than right. The emphasis is on the test rather than on the creation of the hypothesis. This method has been called the "hypothetico-deductive method," the "deductive method of testing," and just "deduction."

The deductive method has also been characterized as "conjectures and refutations," which is the title of a book by Karl Popper (1968b), the most noted philosophical proponent of this view. Popper emphasizes the test aspect and argues that science progresses not by verifying hypotheses but by falsifying them. There is an ultimate truth, but we have no criteria to determine when we have it. We can tell when we do not have it, however, and knowledge can be advanced best by trying to disprove ideas rather than to prove them. The best scientific theory, therefore, is one that has survived repeated vigorous attempts to falsify it.

We may categorize the two aspects of the hypothetico-deductive technique as follows.

## HYPOTHESIS

Although Popper would exclude hypothesis creation from the logical definition of science, it is clear that much time and energy of scientists must be expended in order to develop hypotheses worthy of rigorous test. True genius and creativity in science can be seen in the development of hypotheses that are useful in explanation and prediction and that can be tested in an objective and practical way. The primary criterion for a scientific hypothesis, however, is testability; there must be some way to potentially refute the hypothesis.

Hypotheses vary in the degree to which they include or exclude subject matter; that is, they vary in their explanatory power. Some hypotheses include relatively broad areas and are more general or universal, whereas other hypotheses refer to more restricted areas and are less general or universal in application. Popper uses the term "level of universality" to compare the degree of inclusiveness or generality of hypotheses. More-inclusive hypotheses are formulated at a higher level of universality, and less-inclusive hypotheses are formulated at lower levels of universality. For example, a hypothesis concerning all organisms (e.g., the hypothesis that evolution has occurred) would be formulated at a higher level of universality than one concerning all vertebrates (e.g., the hypothesis that Vertebrata is monophyletic).

All statements, in this view, are hypotheses, even observational or existential ones like "I see a table," or "A table exists here." Such hypotheses are formulated at very low levels of universality and may be considered as "facts." But they must still be falsifiable hypotheses. In the words of Nero Wolfe (in *Gambit*), "I call them facts because if one or more of them can be successfully challenged my theory is untenable."

## TEST

To quote Popper:

> We may if we like distinguish four different lines along which the testing of a theory could be carried out. First there is the logical comparison of the conclusions among themselves, by which the internal consistency of the system is tested. Secondly, there is the investigation of the logical form of the theory, with the object of determining whether it has the character of an empirical or scientific theory, or whether it is, for example, tautological. Thirdly, there is the comparison with other theories, chiefly with the aim of determining whether

the theory would constitute a scientific advance should it survive our various tests. And finally, there is the testing of the theory by way of empirical application of the conclusions which can be derived from it (Popper, 1968a, pp. 32–33).

The last type of test, a comparison of the logical predictions of the hypothesis with observation, is usually the most applicable for phylogeny reconstruction. The tautology test, however, is important for many more general biological hypotheses.

The test aspect of the hypothetico-deductive method stresses the idea that a hypothesis can never be proven correct or true. An idea of hypothesis cannot be confirmed or verified, but it can be falsified by showing it to be inconsistent with observation or experience. When a hypothesis has been tested many times by unsuccessful efforts to falsify it, Popper would say it has been "corroborated." Actually, the strength of such an analysis lies in the falsification of alternative hypotheses. The "corroborated" hypothesis is the one that remains unfalsified.

The falsification of a hypothesis should not be taken as an absolute demonstration that the hypothesis is shown to be wrong for all time. In the hypothetico-deductive system, falsification (or refutation or rejection) means demonstration that the hypothesis being tested (or corollaries or deductions from that hypothesis) is inconsistent with another hypothesis or hypotheses that, for this particular test, are not being questioned. Only hypotheses falsify hypotheses; no statement, test, or observation need be accepted as true by definition. Falsifying hypotheses must in turn be falsifiable, although it is not necessary for them to have already been tested. Usually, falsifying hypotheses are at lower levels of generality than the hypotheses they test. The problem of infinite regress, i.e., that the logical dependence of a statement is continually referred to another statement, does not arise.

The logical falsification of a hypothesis depends only on the potential testability of its apparent falsifier, not on the results of such tests. For any given case of subjecting a particular hypothesis to testing (that is, efforts to falsify it), relevant hypotheses of higher and lower levels of universality take on the appearance of assumptions. In other words, the system deals with one hypothesis at a time.

The results of hypothesis testing, whether rejection or corroboration, do not cause hypotheses to disappear permanently from consideration (in the case of rejection) or to become everlasting fixtures (in the case of corroboration). Nonetheless, Popper would argue that science does genuinely progress in its efforts to explain the real world by this method since new hypotheses often raise new problems. To Popper, the best view of science would be through a study of the problems of science rather than its accumulated data.

Despite the formalized appearance of the hypothetico-deductive model as presented here, the significance of this point of view is not to provide an automatic mechanism for generating "good" science but to provide a means of judging and criticizing ideas. For example, a totally unsatisfactory study can be couched in the jargon of the hypothetico-deductive method and yet make very little contribution to science. In fact, this method makes work harder by requiring us to present our ideas in as explicit a manner as possible and to seek criticism rather than confirmation.

Furthermore, the acquisition of a philosophic framework does not in and of itself make the scientific work done in the framework necessarily better or more understandable. Philosophic jargon may tend to obfuscate rather than clarify methodology. Thus Hull has criticized some philosophers of science for addressing themselves to problems of philosophy rather than problems of science: "No point in logic can be too minute or too artificial to preclude lengthy treatment. Conversely, no fact about the substantive content of science can be so major or so central to science to preclude its dismissal as being beside the point" (Hull, 1975, p. 396).

## Phylogeny Reconstruction and the Hypothetico-Deductive Method

It should be clear that we are interested in a method of testing explicitly stated phylogenetic hypotheses in a manner that will most clearly point out their deficiencies and contradictions. The use of derived character distributions as articulated by Hennig (1966) appears to fit the hypothetico-deductive model best. Bock, who is not a cladist, nonetheless argues that "cladistic methods appear to provide the most severe test by which to attempt to disprove these taxonomic statements" (1973, p. 383). Miles (1973, 1975) also invoked Popperian philosophy as a rationale for testing hypotheses, and his work is distinguished by the use of known taxa rather than the old standbys, A, B, and C. Most importantly, however, as Bonde (1974) has emphasized, Miles (1973, 1975) used derived characters (synapomorphies) to falsify a set of alternative hypotheses, in the manner suggested by Popper, rather than to substantiate a particular hypothesis. "The difference between Miles' approach and that of, e.g., Nelson and Patterson, who use synapomorphies as positive indications of relationships, may seem merely a semantic one—the two methods must in principle lead to the same results—but philosophically it is quite important to recognize testability as more valuable (more 'scientific') than confirmation" (Bonde, 1974, p. 563). Other workers have utilized the hypothetico-deductive system as a rationale for their systematic work, and the reader should refer to Suggestions for Further Reading for examples.

In this section, I will outline a method of phylogeny reconstruction that relies on the hypothetico-deductive system. Beginning with "basic hypotheses," fundamental ideas with high levels of universality, I will proceed to the formulation of phylogenetic hypotheses and methods of testing those hypotheses.

## BASIC HYPOTHESES

If we are going to try to set up a system that is as consistent as possible within the hypothetico-deductive system as developed by Popper, we need to state clearly the most general hypotheses necessary for work in this system. These hypotheses are formulated at the highest level of generality (as far as the study of phylogeny is concerned), and they take on the character of axioms or assumptions with regard to hypotheses of lower levels of generality. However, both of the hypotheses listed below have survived testing for some time and should not be considered as assumptions chosen for expediency.

**Evolution: All the forms of life (i.e., taxa) have been produced by natural processes of inheritance, change, and divergence from a common origin.** As far as I can see, the problem that the evolutionary hypothesis seeks to solve is the reason for the existence of organismic diversity. Historically and logically, the two competing hypotheses are evolution and special creation. The hypothesis that evolution occurred requires that a natural, mechanistic process exist but does not specify the nature of this process. As a corollary of the evolutionary hypothesis, we may state, "Life has one genealogy (phylogeny)." The special-creation hypothesis (as I conceive its rather vague formulations) would allow or require *ad hoc* explanations and multiple genealogies.

**Synapomorphies: New taxa are often characterized by new features.**[2] Whatever hypothesis of evolutionary process one chooses to use, it will have basic units of inheritance, that is, species, populations, gene pools, or whatever. These units are often characterized by evolutionary or genealogic innovations.[3] I am here adopting Hennig's (1966) term "synapomorphy" for such evolutionary novelties.[4] In principle, synapomorphies (or derived characters) can be any detectable attributes of organisms, but morphologic features are the most commonly used in practice.

Our two basic hypotheses, then, are that evolution has occurred and that new taxa may be characterized by new features. Using only these two generalizations, we can develop and test hypotheses about the geometry of descent. There is no reliance upon the "synthetic theory" of evolution or any other particular hypothesis of evolutionary mechanism, and there is no reliance on any particular model or hypothesis of speciation or the nature of species.[5]

---

[2] "New features" also means the loss, obliteration, or transformation of older features. It does not necessarily imply an additional process.

[3] One could substitute "sometimes" or "always" for "often" because the frequency of such innovations is difficult to deal with. If such innovations do not occur when a new species or population originates, it would not be possible to know that this event has occurred, so the statement need not be more specific about frequency.

[4] Hennig (1966) used a complex terminology for his system, and most of his terms have not been utilized by his followers. His terms for the states of characters, however, have been widely used. For reasons discussed later, I would prefer not to use character state analysis, and therefore I do not find his more commonly used terms, such as autapomorphy and plesiomorphy, necessary. The reader, however, will need familiarity with these terms to use the literature; this can be obtained by reading some of the papers referred to in Suggestions for Further Reading.

I also avoid the use of the terms paraphyletic and polyphyletic because of the inconsistencies present in simple definitions and the complexity of consistent definitions (see Platnick, 1977d, and references). I use the term "nonmonophyletic" as a general substitute.

[5] It should be noted that Hennig (1966), Brundin (1966), and some other cladists derived the use of synapomorphies from a dichotomous speciation model. It is not necessary to rely on this

## THE METHOD

**Develop a hypothesis of monophyly.** Phylogeny is the geometric pattern of ancestry and descent among organisms; that is, it is the genealogy of life. Phylogeny reconstruction is the formulation and testing of hypotheses about the geometry of phylogeny. There are two types of hypotheses that can give us useful information about phylogeny:

1. A particular species (or other basic unit of inheritance) is an ancestor of a specified taxon.
2. Two specified taxa have an ancestor in common not found in common with any other taxon.

Much of the current controversy involves the distinctions between these two types of hypotheses. The two types differ in that hypotheses of common ancestry (no. 2) can be tested using only deductions from the two "basic" hypotheses enumerated above, whereas ancestor–descendant hypotheses (no. 1) require additional basic hypotheses or assumptions. Eldredge (in this volume), Tattersall and Eldredge (1977), and Engelmann and Wiley (1977) discuss these additional assumptions, and I would summarize my own interpretation as follows. Ancestor–descendant hypotheses require an assumption concerning the evolutionary mechanism and the nature of inheritance, so that one can test whether or not the taxa used are the reproductive units of evolution,[6] and require also that character reversal does not take place. The first assumption necessitates acceptance of the "synthetic theory" of evolution, or at least one conjunction of its many hypotheses. Although I am

---

hypothesis (or any other) of evolutionary process in order to set up and test phylogenetic hypotheses by cladistic methods. Although some critics (e.g., Mayr, 1974) have mistakenly thought that the method does depend on a speciation model, the error of this viewpoint has been made clear by a number of authors (see especially Platnick, 1977f).

[6] This, I think, is equivalent to Eldredge's assumption (in this volume, pp. 189–90), that "an actual evolutionary event has taken place that does link the taxa at hand." Furthermore, I think that it is equivalent to accepting the objective testability of the "biological species concept" for fossils. (For a useful introduction to problems of species level hypotheses, see Platnick, 1977b.)

hardly arguing that a hypothesis of evolutionary process cannot be formulated within the hypothetico-deductive system, I have not seen such a formulation, and until such a formulation is developed, I will continue to prefer hypotheses that do not rely on particular ideas of the evolutionary process.

The other assumption, namely, that character reversals do not take place, is necessary because the presence of unique features in a hypothesized ancestor is used to refute the hypothesis that such taxa are ancestors; if character reversals do occur, it is no longer possible to test ancestor hypotheses with characters. The problem with this hypothesis is not so much that it is unfalsifiable but that it does seem to have been falsified. There are some phylogenetic hypotheses that require *ad hoc* invocations of character reversal in order to be consistent with the hypothesis enumerated earlier, that evolution has taken place.

This brings us to the second major distinction between hypotheses of ancestry and descent versus common ancestry; namely, what is the information content of such hypotheses and what further ideas do they lead to? Hypotheses of common ancestry include hypotheses of direct ancestry. That is, in the statement "X and Y have an ancestor in common not found in common with any other taxon," X could be the ancestor of Y or vice versa. This type of information is left unspecified. What do we gain by making ancestor–descendant hypotheses? Eldredge (in this volume, p. 190) says, "It takes a tree—an explicit, specific hypothesis of ancestry and descent—to provide fuel to the evolutionary fire." Perhaps this is true, but we have had specific examples of ancestor–descendant hypotheses for some time. How have these examples aided hypotheses about evolutionary process and mechanism? One example might be the current gradualism-versus-punctuated-equilibrium controversy. Specific hypotheses of ancestry and descent should be able to resolve this problem, but they have not. Why not? Because such hypotheses are so dependent on the very answers to the questions they seek to solve that any example can be interpreted as being consistent with either speciation hypothesis. Again, this is due to the reliance of ancestor–descendant hypotheses on hypotheses concerning the evolutionary mechanism and the nature of species.

In summary, then, the emphasis in phylogeny reconstruction should be on hypotheses of common descent because these hypotheses are less dependent on auxiliary hypotheses that are themselves unfalsifiable, false, or inadequately tested at present.

A hypothesis of common descent is the same as a hypothesis that a particular taxon is monophyletic. A taxon is monophyletic when all of its members have an ancestor in common that is not common to any other known taxon. The simplest form of such a hypothesis is the statement, "Two taxa have an ancestor in common not found in common with any other taxon."[7] Hypotheses of monophyly may be represented diagrammatically, for convenience, as cladograms, a device most fully developed by Hennig (1966). Cladograms are not intended to be phylogenetic trees in the strict sense in which each line represents a lineage and each node represents a speciation event. The lines on a cladogram do not necessarily represent evolutionary lineages, and the length, angle, and spacing of lines means nothing. Each pair of connected lines may be rotated at the connecting node with no change in meaning.

A cladogram, then, neither specifies ancestor–descendant relationships nor precludes them. For example, in figure 4A, discussed later, taxon Y could be an ancestor of Z, and the cladogram, as a hypothesis of monophyly, would still be an accurate statement (i.e., that Y and Z have an ancestor in common not shared by X).

In a cladogram, a series of taxa are chosen to be the subject of a hypothesis of relationship. The taxa used as the smallest units in any particular hypothesis of this sort are "basic taxa." These taxa have one important property: they are in themselves hypotheses of monophyly at a lower level of universality. They may also represent hypotheses about the process of evolution, that is, they may be species or populations, but in any case, they must be falsifiably monophyletic. Philosophically, basic taxa act as individuals in any particular cladogram.[8] For the purposes of any one particular cladogram, it is not necessary logically that the monophyly of the basic taxa be already tested. All that is necessary for this method to be consistent with the hypothetico-deductive system is that the monophyletic nature of the basic taxa be potentially falsifiable.

Hypotheses with many taxa included in them are all resolvable into a series of statements of the form "Two taxa have an ancestor in common not found in common with any other taxon." This is the most generalized form of the "three-taxon statement." In order to facilitate testing, we may specify

---

[7] The hypothesis can also be stated to read "Two *or more* taxa have an ancestor in common . . ." but *at least* two taxa must be involved in a statement of the hypothesis, as well as the third "taxon" constituted by all other organisms.

[8] Ghiselin (1974) discusses the philosophical applicability of the terms "individual" and "class" with regard to systematic taxa, species in particular. As Bonde (1977, p. 757) has remarked, "Logically a clade, just like a species, can be treated as an individual." Actually, any of the things I call "basic taxa" are treated as individuals within the hypothesis being tested.

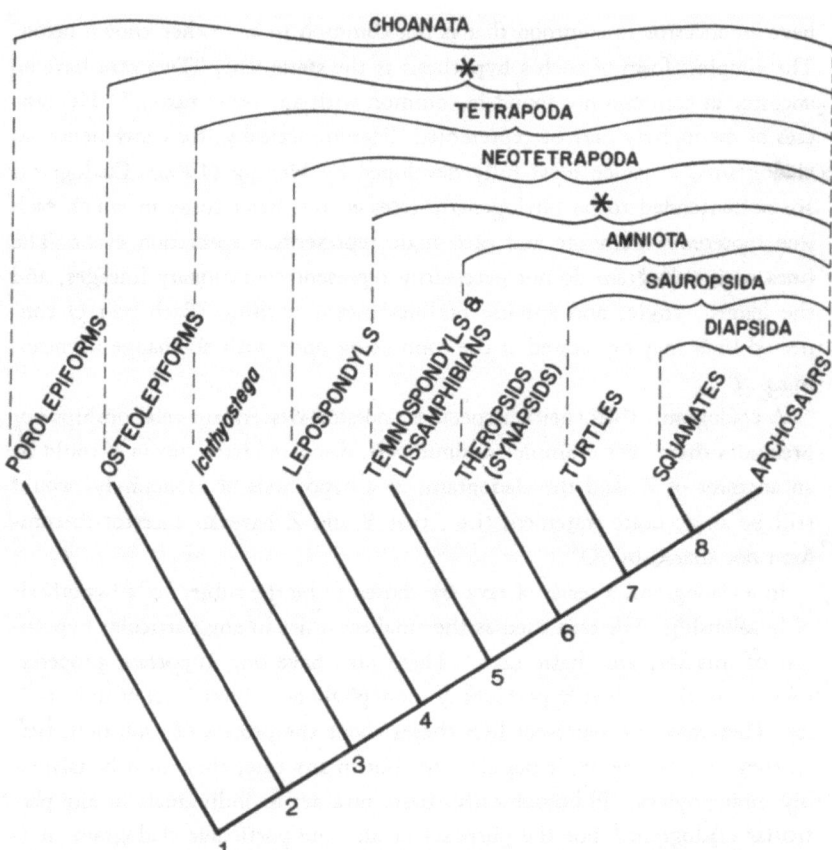

Figure 1. A theory of relationships of the tetrapods and rhipidistian fishes. The numbers refer to synapomorphies possessed by the taxa above the nodes. Table 1 lists these characters. The two asterisks are as yet unnamed groups.

the third taxon rather than say "all other taxa," and we may specify that the three taxa together form a monophyletic group (see the section on testing for discussion of this aspect). In any case, all phylogenies consist of a series of internesting three-taxon statements. Figures 1 and 2 show how a phylogenetic

Figure 2. The theory of tetrapod relationships seen in figure 1, resolved into a series of internesting three-taxon statements in order to show the logical structure of the cladogram. The arrows indicate three-taxon hypotheses included in the preceding three-taxon statement. The numbers refer to synapomorphies seen in figure 1 and table 1. Porolepiforms are deleted for simplicity.

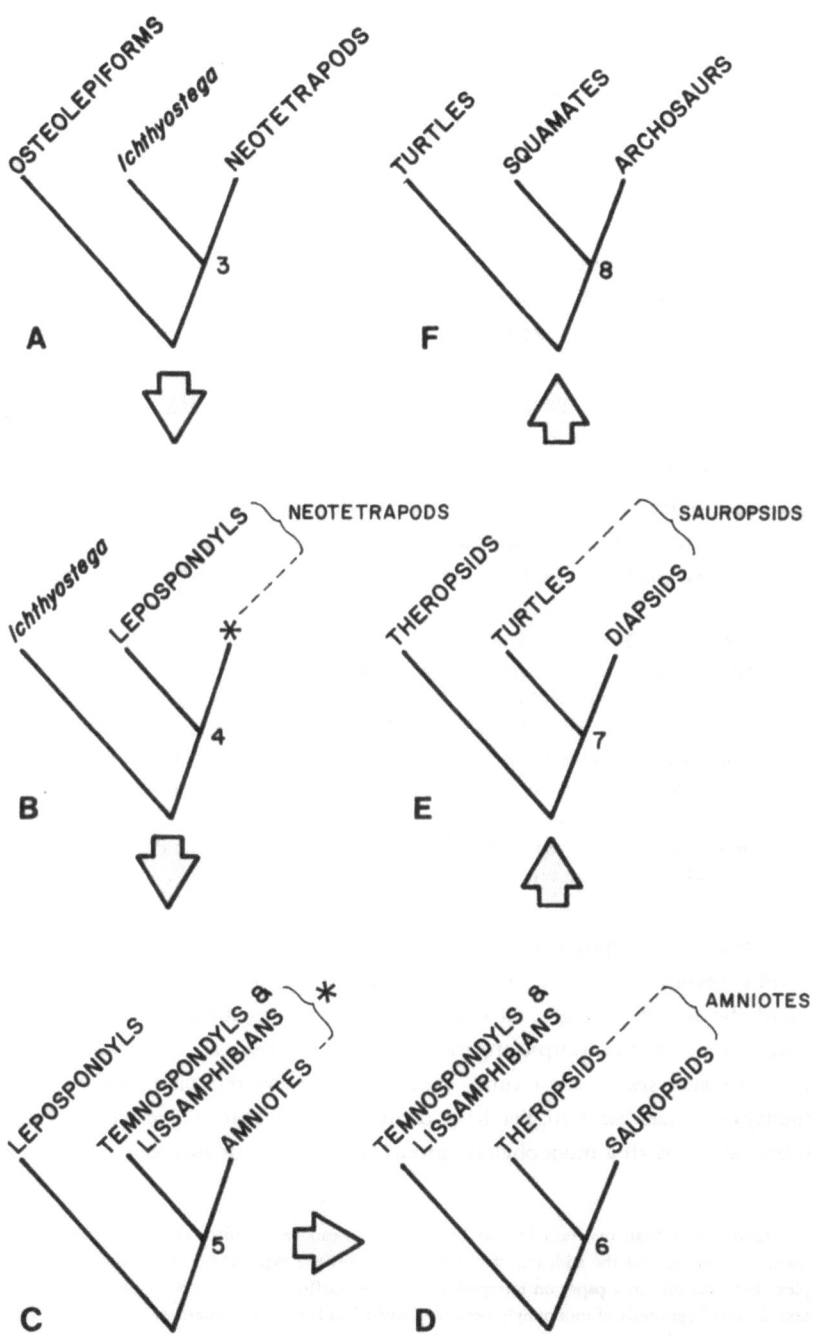

hypothesis about a number of taxa in the Choanata really consists of a series of three-taxon hypotheses, each hypothesis having a higher or lower level of universality than the next and each one being tested by its own series of synapomorphies (see also table 1).

**Test the hypothesis of monophyly with synapomorphy (shared derived character) distributions.** Keeping in mind the two axioms enumerated earlier, we may attempt to falsify a hypothesis set up by the criteria just discussed. That is, we may test a hypothesis of monophyly by examining the involved taxa for synapomorphies because we expect the members of a monophyletic group to possess one or more unique characters in common. Platnick and Cameron (1977) present an interesting discussion of the universal applicability of analyses of this sort, showing that areas outside of biology also utilize techniques of a similar logical structure. In terms of the hypothetico-deductive system, we have the following:

> *Hypothesis:* Two taxa, A and B (the basic taxa of this hypothesis) have an ancestor in common not found in common with any other taxon.[9]
> *Tests:* (1) Search for characters (synapomorphies) that are found in A and any other taxon except B, and characters found in B and any other taxon except A. If a character is found with either of these distribution patterns, then, in principle, it falsifies this hypothesis of monophyly. (2) Search for characters (synapomorphies) in common between A and B but not any other taxon. This falsifies alternative hypotheses of the form "A plus any other taxon (except B) is monophyletic" and "B plus any other taxon (except A) is monophyletic."

In testing monophyletic groups with distributions of synapomorphies, it rapidly becomes apparent that a shortcut is available that will allow us to avoid the search through "all other taxa" referred to above. This procedure may be termed "outgroup comparison," and it has been an important part of character analyses for some time. In this method, we rely on a hypothesis of monophyly that has a higher level of universality; that is, we replace "any other taxon" with a monophyletic group larger than but including A and B.

---

[9] Again, more than two taxa (A and B in this case) can be specified, but two is the lowest number possible, and the addition of others only makes the explanation of testing more complex. For example, in a paper on tetrapod monophyly (Gaffney, 1979), I chose to use five basic taxa for one hypothesis of monophyly because I wished to have as few auxiliary phylogenetic hypotheses as possible.

Table 1. Shared derived characters involved in tests of the phylogenetic hypothesis. The numbers below correspond to the numbers in figures 1 and 2. The references allow the reader to find figures and descriptions of the characters. Porolepiforms are excluded, as in figure 2.

1. Choanata (Miles, 1975; see also Nelson, 1969; Säve-Söderbergh, 1934).
   a. Bipartite braincase ossifications (Moy-Thomas and Miles, 1971).
   b. Internal choanae (Andrews, 1973; Panchen, 1967).
2. Unnamed taxon.*
   a. Distinctive humeral morphology (Andrews and Westoll, 1970a,b).
   b. Dorsal ribs (Andrews, 1973).
3. Tetrapoda (Goodrich, 1930).
   a. Development of space between skull and shoulder girdle (Andrews and Westoll, 1970a).
   b. Development of distinctive carpus, tarsus, and pentadactyly (Andrews and Westoll, 1970a; Gregory, 1915, 1949; Schaeffer, 1965).
   c. Illium connected to vertebrae by ossified ribs (Andrews and Westoll, 1970a).
   d. At least partial braincase fusion (Jarvik, 1955).
   e. Loss of dermal roof joint (Thomson, 1967).
   f. Reduction of preopercular and opercular bones (Jarvik, 1955).
   g. Posterior movement and elongation of parietals and frontals, reduction of postparietals (Schaeffer, 1965; Westoll, 1943).
   h. Middle ear with distinctive stapes (Thomson, 1966).
4. Neotetrapoda.
   a. Persistent notochord not entering braincase (Jarvik, 1955).
   b. Preopercular and opercular absent (Jarvik, 1955; Schaeffer, 1965).
   c. Median bony fin supports gone (Jarvik, 1952).
5. Unnamed taxon.*
   a. Bipartite vertebrae (Romer, 1947). (It should be noted that this is one of the weakest aspects of this phylogeny.)
6. Amniota (Goodrich, 1930; Wilder, 1887).
   a. Pterygoid flange (Carroll, 1969, 1970).
   b. Astragalus (Carroll, 1969, 1970).
   c. Closed otic notch (Parrington, 1958).
7. Sauropsida (Goodrich, 1916, 1930).
   a. Aortic arch morphology (Goodrich, 1916).
8. Diapsida (Romer, 1956).
   a. Diapsid skull fenestration (Romer, 1956).
   b. Prominent palatine fenestrae (Romer, 1956).

For example, in developing tests of the various hypotheses presented in figure 1, we may accept monophyly of all Vertebrata, and in searching for characters unique to Tetrapoda, to take an example, we will limit our search to Vertebrata. Actually, however, the best method is probably to limit the size of the higher-level hypothesis as much as possible in order to reduce mistakes in character recognition. For example, in testing monophyly of Sauropsida (fig. 1), it would be best to assume monophyly of Amniota and search for distributions of characters among turtles, Diapsida, and the remaining Amniota.

In summary, we may rely on a higher-level hypothesis of monophyly in order to ease our search for falsifying characters. Although, in principle, the search for characters is throughout all known taxa, we may limit our search to a more restricted "outgroup" based on the results of previous efforts. If these previous efforts are wrong, then the tests based on those hypotheses may also be wrong, but, contrary to Nelson (1973a), there is no logical dependence on "higher-level phylogenies," because, in principle, each hypothesis is falsifiable independently of all other phylogenetic hypotheses by comparison with all other organisms.[10]

Another way of restating the above method is to assume that the group $X + Y + Z$ is monophyletic and that each taxon is in itself monophyletic. In this case, there are three possible alternative hypotheses (see fig. 3).[11] Testing these with character distributions will result in either all three being falsified to some degree or the emergence of one hypothesis as unfalsified or "least rejected" (Wiley, 1976). Even if all three are rejected at least once, the least-rejected hypothesis should be chosen for further testing and incorporation into the larger system of phylogenetic hypotheses. It is a corollary of the system that one of the hypotheses survives, even if it has been falsified at least once (see Wiley, 1976, for a good example of this). This is a result of circumscribing the possible number of hypotheses and recognizing that par-

[10] One might argue that this is dependent on the monophyly of life, but it seems to me that it is dependent only on the basic hypotheses presented earlier (i.e., evolution and synapomorphies).

[11] A fourth possibility, a trichotomy, is not really a new hypothesis, because it only restates the assumption of $X + Y + Z$ monophyly. If we apply tests to the three dichotomous possibilities and each is falsified an equal number of times, then no choice can be made among them, but this does not really indicate that a trichotomy is a corroborated hypothesis. A trichotomy may nonetheless appear in a hypothesis (as in Gaffney, 1975) owing to the lack of a clear result in testing the alternative dichotomous hypotheses.

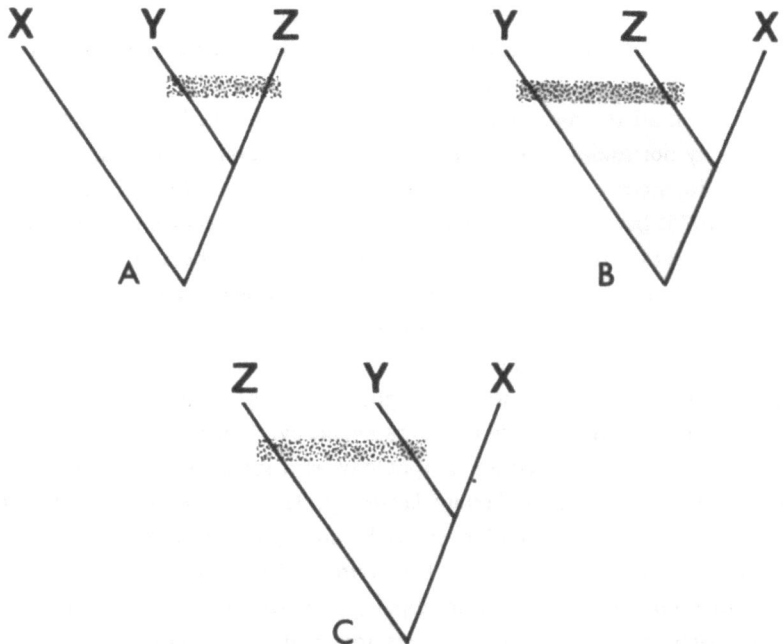

Figure 3. An example of a synapomorphy distribution (represented by the bars) falsifying two out of three alternative hypotheses relating three taxa, X, Y, and Z. In this case, hypotheses B and C are falsified, whereas hypothesis A is not.

allelism exists,[12] thereby saving one of the hypotheses by the application of *ad hoc* hypotheses of parallelism.

## SYNAPOMORPHY AND HOMOLOGY

Phylogenetic hypotheses are tested by character distributions, but according to the hypothetico-deductive system, all tests are in themselves falsifiable

[12] Throughout this paper, I use the word "parallelism" to include also what others call convergence and reversal. The distinctions among these three concepts require hypotheses about the evolutionary mechanism at a level of analysis that I am not dealing with here.

hypotheses. How are characters, i.e., synapomorphies, tested?[13] The hypothesis that a character is a synapomorphy may be stated as follows: "An attribute with a stated series of specifications may be found, modified or unmodified, in all the members of a specified group." The hypothesis may be falsified by not finding the specifications stated in examinations of topology and development (in the case of morphologic features). In other words, as Wiley (1975) has argued, the criteria for testing synapomorphy are the same as those traditionally used for testing homology. This is because both are hypotheses about common ancestry. Both words, however, refer to hypotheses of common ancestry at different levels of universality. For example, when we say that the forelimb of frogs and bats is homologous, we are actually arguing that both taxa belong to a monophyletic group (Tetrapoda) which has forelimbs of a particular sort as a synapomorphy. Conversely, an argument proposing the use of forelimb morphology as a synapomorphy testing the monophyletic nature of the Tetrapoda (see fig. 1) is also an argument that all tetrapod forelimbs are homologous. A homology at one level is a synapomorphy at some higher level. All of the criteria classically used for homology determination, i.e., development, topology, morphology, the basic features of comparative biology, are also tests for monophyly. This makes sense, because phylogenetic relationship is the fundamental comparative feature of organisms.

## CHARACTER CONTRADICTIONS, PARSIMONY, AND PARALLELISM

When multiple series of characters are used, it is often the case that all three possible hypotheses relating three taxa will appear to be falsified by one or more character distributions (fig. 4). In this case, the least-rejected hy-

[13] Some cladists (e.g., Patterson, 1978) argue that synapomorphy identification is not a hypothesis but a "discovery" of reality and that hypotheses of monophyly are not falsifiable, because they are "singular" statements about unique events. In this paper, I am presenting the notion that Popper's view of the hypothetico-deductive system will improve our efforts at phylogeny reconstruction by clearly showing that all aspects are open to criticism (falsification) and no statement of authority need be accepted. If one systematist "discovers" a synapomorphy that contradicts another systematist's "discovery," how can we rationally choose one over the other? The importance of cladism is its compatibility with the hypothetico-deductive method; a reversion to authoritarianism is a step backward.

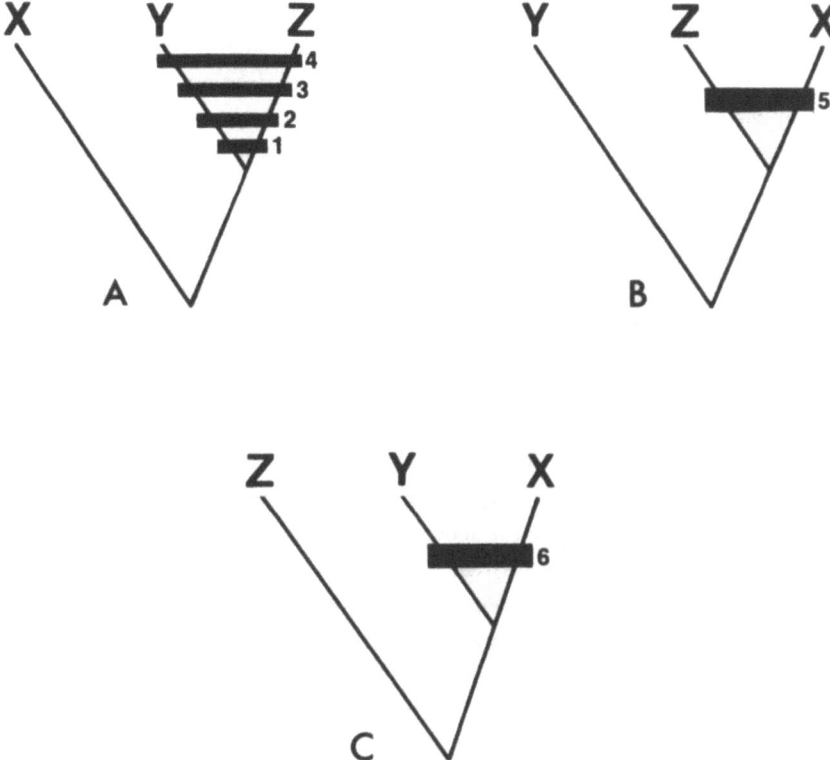

Figure 4. Three alternative hypotheses relating the same three taxa, X, Y, and Z. This example illustrates a situation in which each alternative has been falsified by at least two apparently "correct," shared derived characters. That is, hypothesis A is falsified by characters 5 and 6, hypothesis B is falsified by characters 1–4 and 6, and hypothesis C is falsified by characters 1–5. Hypothesis A is to be preferred for further testing and biologic use because it is the least-rejected hypothesis; that is, it requires the fewest *ad hoc* hypotheses for it to be true. The preference for hypothesis A requires that characters 5 and 6 be interpreted as parallelism.

pothesis (the one falsified the fewest number of times) is the one incorporated into the interresting system of hypotheses and subjected to further testing. The principle of parsimony is widely invoked as the basis for this decision.

One of the most basic ideas in the history of scientific explanation is that often called parsimony or Ockham's razor. This idea is usually stated as follows: Among two or more conflicting solutions to a given problem, the simplest solution (i.e., the one involving the smallest number of logical steps or auxiliary conditions) should be chosen, all other factors being equal. In this sense, parsimony might be considered a methodological rule, invoked only because to do otherwise could result in no choices being made and no progress possible, not because it mirrors reality in some way.[14] In practical terms, the use of parsimony enters into almost every decision we make. More complicated solutions to problems often involve greater expenditure of resources or time, and in situations of this sort, it hardly seems to be an esoteric, arbitrarily chosen rule of logic. However, the original statement of Ockham's razor appears to be, "We are not allowed to affirm a statement to be true or to maintain that a certain thing exists, unless we are forced to do so either by its self-evidence or by revelation or by experience or by a logical deduction from either a revealed truth[15] or a proposition verified by observation" (Boehner, 1957, p. xx).[16] And also, "We must not affirm that something is necessarily required for the explanation of an effect, if we are not led to this by a reason proceeding either from a truth known by itself or from an experience that is certain" (ibid. p. xx). From this, it seems to me that parsimony, or Ockham's razor, is equivalent to "logic" or "reason" because any method that does not follow the above principle would be incompatible with any kind of predictive or consistent system.

In any case, in a hypothetico-deductive system, parsimony is not merely a methodological convention; it is a direct corollary of the falsification criterion for hypotheses (Popper, 1968a, pp. 144–45). When we accept the hypothetico-deductive system as a basis for phylogeny reconstruction, we try to test a series of phylogenetic hypotheses in the manner indicated above. If all three of the three possible three-taxon statements are falsified at least once, the least-rejected hypothesis remains as the preferred one, not because of an arbitrary methodological rule, but because it best meets our criterion of testability. In order to accept a hypothesis that has been successfully falsified one or more times, we must adopt an *ad hoc* hypothesis for each falsification. An *ad hoc* hypothesis is required "for the sole purpose of saving a hypothesis seriously threatened by adverse evidence; it would not be called for by other findings, and roughly speaking, it leads to no additional test implications" (Hempel, 1966, p. 29). Therefore, in a system that seeks to maximize vul-

---

[14] Strictly speaking, if the rule is used only to provide a means of escaping a dead end, then it could just as well choose the most complex alternative.

[15] It should be realized that Ockham was a Catholic philosopher of the Middle Ages and worked within a context that required that ultimate causes be explained by Holy Writ.

[16] I am indebted to Mr. William Kohlberger for bringing this source to my attention and allowing me the free use of his unpublished manuscript on this topic.

nerability to criticism, the addition of *ad hoc* hypotheses must be kept to a minimum to meet this criterion. However, as Popper remarks, "We need not reject . . . every auxiliary hypothesis that fails to satisfy these standards" (1968a, p. 83), because in a three-taxon test, as discussed earlier, one of the three alternatives is "true" (given the required assumption of monophyly of $X + Y + Z$), and therefore, if all three alternatives have been falsified, at least one of the *ad hoc* hypotheses is also "true." So, contrary to Hempel's statement above, the *ad hoc* hypotheses do lead to further test implications, and we should make efforts to bring the *ad hoc* statements out of the untestable category and try to deal with them in an empirical fashion. Unfortunately, I do not see how this can be done in practice at the present time.

In phylogeny reconstruction, the most common *ad hoc* hypothesis in an instance of character contradiction is the appeal to parallelism. Often a parallelism argument will appear before a hypothesis of relationship is even tested, with the idea that such an argument has some merit at this stage of analysis. But an appeal to parallelism or convergence alone cannot be used to test any theory of relationships. What criteria exist that allow an identification of an example of parallelism or convergence? "A judgment of convergence must be based on an *a priori* assumption of relationship. The fact that two taxa . . . might show a number of morphological differences cannot serve as an argument for non-relationship. The latter can only be proposed once a relationship has been demonstrated between one of these taxa and a third taxon" (Cracraft, 1972, p. 387). An argument or statement of parallelism is always reducible to an argument in favor of one phylogeny over an alternative one. Furthermore, I would say that the only objective evidence we have for the existence of parallelism, convergence, and reversal is the persistent appearance of character contradictions (see also Platnick, 1977a).

It has been argued that choosing a hypothesis on the basis of parsimony, that is, choosing a least-rejected hypothesis, is really no choice at all, because each falsification is based on a character distribution, and one may object to the notion that two or more characters are more "important" or have more "weight" than one character. A recent, explicit discussion of this problem has been presented by Hecht and Edwards:

> Realizing that his methodology might allow hidden parallelisms, Heyer . . . thought . . . that the use of great numbers of characters would override the few

mistaken parallelisms. They all fail to realize that, in the recognition of branching points in a phylogeny, only one reliable character is necessary to indicate lineage. The clouding of the pattern of relationships by unnecessary data prevents the recognition of the real lineages. . . . From the viewpoint of lineage detection, it is more important to use a few well-analyzed morphoclines than many poorly or improperly analyzed ones (1977, p. 13).

This last notion I would agree with wholeheartedly, but I would interpret the statement as referring to synapomorphy distributions which omit important taxa, have ambiguous criteria for their identification, or suffer from other problems of lower-level hypothesis formulation and test. To Hecht and Edwards, "analysis" refers to the *a priori* test of characters for examples of parallelism and the classification of these characters into weighted categories.

This *a priori* test of characters requires a reliance on a hypothesis of the evolutionary mechanism. Even if a well-tested hypothesis of this sort existed, it would be preferable to maintain independence between it and a system for genealogy reconstruction, so that each could be used to test the other. In order to weigh characters *a priori* in schemes such as that of Hecht and Edwards, there must be a method of relating morphologic features (or "functional complexes" or whatever) to evolutionary mechanism hypotheses, such as natural selection. At present, I see no objective criteria to allow us to identify or associate a "selective force" (or some similar idea) with some circumscribed unit of morphology. Even if this were possible, the result would be the comparison of selective forces, with the implication that each force is of a known magnitude with respect to the others and that this allows a choice to be made. Again, I see no objective criteria that can be used to weigh selective forces; the whole idea of comparing morphologic units on this basis is wishful thinking. Although the delimitation of characters and their use as single units with some sort of implied equivalence is an important problem in cladism (as in every method of systematics), alternative methods requiring information on the evolutionary mechanism do not solve the problem, but simply refer it to another level of hypothesis testing. Furthermore, it would seem to be much more difficult to deal with *ad hoc* hypothesis identification and resolution at the level of evolutionary mechanism hypotheses, because this area is already riddled with untestable hypotheses.

As developed earlier, hypotheses may be arranged in levels of universality. The arrangement of *ad hoc* hypotheses in this fashion brings up an important factor in the use of *a priori* judgments of character dependability. If, for example, we have a character contradiction that requires one or another *ad hoc* hypothesis for a phylogenetic hypothesis to remain as a viable alternative, the logical nature of the *ad hoc* hypotheses provides the basis for a choice among them. If, in this example, we can explain the feature by a hypothesis of relatively low universality, say a single instance of parallelism, this is to be preferred to a hypothesis having a relatively high level of universality, say the adoption of general criteria for the *a priori* weighting of characters. General hypotheses of parallelism must be formulated such that they are applicable to all organisms; they have a high level of universality and thereby weaken the whole system. Lower-level *ad hoc* hypotheses, even though numerous, are restricted in scope, are not necessarily required for each phylogenetic hypothesis, and are therefore to be preferred. Once again, all of this depends on agreement with my statement that *a priori* hypotheses of parallelism are untestable. If someone can formulate testable hypotheses in this area, then my argument is negated, but they will have to formulate testable hypotheses for a great deal of the "synthetic theory of evolution" to accomplish this (Platnick, 1977e).

## MORPHOCLINES AND CHARACTER STATES

Hennig (1966) and most cladists (see especially, Hecht and Edwards, 1977; Schaeffer, Hecht, and Eldredge, 1972) emphasize the use of morphoclines and character states as methods of character analysis. There are, however, distinct problems with this approach. This method consists of the arrangement of different variations of morphology, which are presumed to be homologous at some higher level, into a linear sequence (morphocline). By outgroup comparison, one end of the sequence is hypothesized as primitive (plesiomorphous) and one end as derived (apomorphous). The morphocline, or character phylogeny, is then applied as a test to a cladogram. The whole morphocline is usually considered to be a character, with the subdivided variations of it the character states, some relatively plesiomorphic, some relatively apomorphic. I conclude that, in many cases, this method is the same as the one I have stated above as test 2 ("Search for characters in common between A and B but not any other taxon"), but where it differs from that method, it is logically invalid.

To begin with, there are "character phylogenies," but they can only be deduced by tracing a pattern of structures through a preexistent phylogenetic hypothesis, and are a corollary of that hypothesis. Morphoclines consisting of only two character "states," one plesiomorphic and one synapomorphic, can be translated to read the same as the tests discussed earlier. That is, the synapomorphic "state" is simply a character unique to a group. Identifying a plesiomorphic "state" is logically unnecessary for this test and only complicates matters by referring to a phylogeny formulated at a higher level of universality (more inclusive phylogeny) at which the plesiomorphic "state" is a synapomorphy. Whether we call these features "states" or "characters" matters little in this case. The invalid use of morphoclines occurs when more than two "states" are hypothesized. In a particular case, a morphocline might have four states, a–b–c–d, and the standard analysis would be to decide whether the a or the d end is synapomorphic. But what about the alternative sequences, b–d–c–a, c–b–a–d, etc.? Even before the polarity of the morphocline is sought, the statement of it precludes large numbers of alternative sequences. How have these alternatives been rejected? Ontogenetic sequences might be used where available, but this would require out-group comparison of ontogenies and would not differ fundamentally from the type of tests referred to above, in that it would be resolvable into separate searches for the presence or absence of features in taxa. In any case, the sequences chosen for multistate morphoclines usually seem to be based on implicit phylogenies or the supposed recognition of evolutionary phenomena such as the simplification of structures, size increase or decrease, or some other process.

Finally, morphocline analyses are usually presented as being formulated and tested independently of phylogenetic hypotheses, in that one end of a sequence is judged plesiomorphic and the other end apomorphic without reference to any particular three-taxon statement. However, a character can be considered synapomorphic for only one phylogenetic hypothesis, i.e., one monophyletic group; it is not a synapomorphy within the group (it is a plesiomorphy) and does not exist outside the group. Morphocline analysis can result in the application of characters at the wrong level of universality with regard to phylogenetic hypotheses.

In practice, morphocline analysis and the testing technique I advocate here will usually have the same results, particularly in the case of two-state morphoclines. But morphoclines of more than two states include implicit hypotheses that are logically invalid. In any case, in spite of the traditional association of morphoclines and cladism, the use of morphoclines has no advantage over the type of testing described above and should be avoided.

## CLASSIFICATION: STABILITY IS IGNORANCE

The furor over classification is seemingly endless. The discouraging part of it is that most of the discussion concerns sociological problems in science rather than substantive biological problems. Many who agree on the geometry of a phylogenetic hypothesis will disagree strongly on the purposes and implications of producing a classification. There is a strong traditional feeling that stability of some sort is important in classification, especially in regard to nomenclature. But when classifications are presented as biological hypotheses, then we must question the usefulness of stability. In the literature, a classification is often the most important, if not the only, way of presenting phylogenetic hypotheses and conclusions of other systematic work. This is unfortunate because the formulation of scientific ideas in a context that imposes stability for traditional purposes breeds the illusion that stability of classifications demonstrates accuracy and depth of understanding. In fact, temporal stability of classifications often reflects ignorance of relationships and lack of work. I hardly advocate change for its own sake, but the maintenance of names for discarded concepts seems useless and misleading.

One often hears of a "cladistic classification" (Mayr, 1974), but so-called cladists use different methods for translating their phylogenetic hypotheses into words (see Bonde, 1977, for a review of this topic). Hennig (1966) advocates a system using geologic age as the rank determinant, Crowson (1970a) would classify Recent and fossil groups in separate classifications, and Nelson (1969, 1972a, 1973b) differentiates fossils from Recent forms by sequencing the former and ranking the latter. The most straightforward method of reflecting a phylogenetic hypothesis is that advocated by Schaeffer, Hecht, and Eldredge (1972) and demonstrated by McKenna (1975). In this method, all taxa are treated alike, and the classification is an exact, redundant, image of the cladogram.

Although there have been strong arguments to include other sorts of biological information in a classification ("genetic content," Mayr, 1974; "patristic," or primitive, characters, Simpson, 1975), it seems to me that the most important biological reference system is a phylogeny and that a classification that mirrors a phylogenetic hypothesis (however transitory) is the most useful for systematists and nonsystematists. However, as long as there are nonscientific (nontestable) elements (tradition, stability, authoritarianism) that operate in the development of a classification, such classifications will continue to mislead systematists and nonsystematists alike.

# Acknowledgments

Dr. Gareth Nelson has provided the impetus for much of the current interest in systematic methodology, and many individuals, myself included, are indebted to him for sharing his originality and point of view.

I have been dependent on my colleagues and the literature for most of the ideas expressed in this paper. The following individuals particularly aided the progress of the manuscript, either by reading it or by discussing aspects of the problem: S. Barghoorn, D. Chure, R. Coldiron, N. Eldredge, W. Kohlberger, M. McKenna, R. Miles, G. Nelson, N. Platnick, D. Rosen, B. Schaeffer, and T. Schuh. I am especially grateful to George Engelmann for letting me use some of his unpublished work and for discussing most of the points in this paper.

# Suggestions for Further Reading

The past 5 years have seen a distinct increase in the number of papers dealing with cladistic methodology. Hennig (1950), and the rewritten English version (1966), are the primary sources for the most fully developed treatment of phylogenetic systematics, but as Bonde (1977) has documented, there is a great diversity of ideas and methods currently referred to as cladism or phylogenetic systematics.

As a beginning foray into the burgeoning literature on cladism, I would suggest reading Bonde (1977), Hennig (1965, a summary paper), and Wiley (1976, pp. 7–13). The following categories may act as a guide to further literature.

**Papers of general interest with regard to phylogenetic systematics.** Bonde, 1974, 1977; Brundin, 1966, 1968; Cracraft, 1972, 1974; Crowson, 1970a (systematics text, see reviews by Nelson, 1971b; Rosen, 1974a); Eldredge and Tattersall, 1975; Griffiths, 1972, 1974; Hennig, 1965, 1966; Løvtrup, 1977; Nelson, 1970, 1971a, 1972a, 1972b, 1973a, 1973b; Platnick, 1977a, 1977d, 1977f; Platnick and Cameron, 1977; Schaeffer, Hecht, and Eldredge, 1972; Wiley, 1976.

**Philosophic aspects of systematics.** The primary references for the Popperian view of the hypothetico-deductive system are Popper (1968a, b) and, of these, chapters I through VII and X of the former are the most pertinent. A good way to begin reading in this area is through Magee (1973, 1974) and Platnick and Gaffney (1977, 1978a,b). The latter are intended expressly for the purpose of informing systematists about relevant ideas in Popper's works.

Other works relevant to philosophical problems in systematics are: Ackermann, 1976; Hempel, 1966; Hull, 1970, 1974, 1975; Kuhn, 1970; Lakatos and Musgrave, 1970; Popper, 1960, 1972, 1976 (these three are reviewed by Platnick and Gaffney, 1978a); Schilpp, 1974; Suppe, 1977.

**Criticisms of phylogenetic systematics.** Ashlock, 1974; Bock, 1968, 1973 (see Wiley, 1975); Campbell, 1975; Darlington, 1970 (see Nelson, 1971a; Brundin, 1972), 1972; Harper, 1976; Hecht and Edwards, 1977; Mayr, 1974 (probably the most lucid and sophisticated criticism, see also rebuttals by Hennig, 1975; Nelson, 1974; Rosen, 1974b); Simpson, 1975; Sokal, 1975; Szalay, 1977.

**Examples of phylogenetic systematics.** Brundin, 1966; Eldredge, 1972, 1973; Eldredge and Tattersall, 1975; Gaffney, 1975, 1977, 1979;

Griffiths, 1972; Hennig, 1969 (see review by Crowson, 1970b); MacFadden, 1976; Nelson, 1968, 1969, 1973c; Platnick, 1977c; Platnick and Gertsch, 1976; Rosen, 1973, 1974c; Wiley, 1976.

## References

Ackermann, R. J. 1976. *The Philosophy of Karl Popper*. University of Massachusetts Press, Amherst.
Andrews, S. M. 1973. Interrelationships of crossopterygians. In P. H. Greenwood, R. S. Miles, and C. Patterson (eds.), *Interrelationship of Fishes*, pp. 137–77. Academic Press, London.
Andrews, S. M., and Westoll, T. S. 1970a. The postcranial skeleton of *Eusthenopteron foordi* Whiteaves. *Trans. Roy. Soc. Edinburgh* 68:208–329.
Andrews, S. M., and Westoll, T. S. 1970b. The postcranial skeleton of rhipidistian fishes excluding *Eusthenopteron*. *Trans. Roy. Soc. Edinburgh* 68:391–489.
Ashlock, P. D. 1974. The use of cladistics. *Ann. Rev. Ecol. Syst.* 5:81–99.
Bock, W. J. 1968. Phylogenetic systematics, cladistics and evolution. *Evolution* 22:646–48.
Bock, W. J. 1973. Philosophical foundations of classical evolutionary classification. *Syst. Zool.* 22:375–92.
Boehner, P., tr. 1957. *Ockham—philosophical writings*. Bobbs-Merrill, Indianapolis, Ind.
Bonde, N. 1974. Review of "Interrelationship of Fishes." *Syst. Zool.* 23:562–69.
Bonde, N. 1977. Cladistic classification as applied to vertebrates. In M. Hecht, Goody, P. C. and B. M. Hecht (eds.), *Major Patterns in Vertebrate Evolution*, pp. 741–804. NATO Advanced Study Institutes Series. Plenum Press, New York.
Brundin, L. 1966. Transantarctic relationships and their significance, as evidenced by chironomid midges with a monograph of the subfamilies Pondonominae and Aphroteniinae and the austral Heptagyinae. *K. Svenska Vetenskapsakad. Handl.* 11:1–472.
Brundin, L. 1968. Application of phylogenetic principles in systematics and evolutionary theory. In T. Ørvig (ed.), *Current Problems of Lower Vertebrate Phylogeny* pp. 473–95. Interscience, New York.
Brundin, L. 1972. Phylogenetics and biogeography. *Syst. Zool.* 21:69–79.
Campbell, K. S. W. 1975. Cladism and phacopid trilobites. *Alcheringa* 1:87–96.
Carroll, R. L. 1969. Problems of the origin of reptiles. *Bio. Rev.* 44:393–432.
Carroll, R. L. 1970. The ancestry of reptiles. *Phil. Trans. Roy. Soc. London* 257:267–308.
Cracraft, J. 1972. The relationships of the higher taxa of birds: Problems in phylogenetic reasoning. *Condor* 74:379–92.

Cracraft, J. 1974. Phylogenetic models and classification. *Syst. Zool.* 23:71–90.
Crowson, R. A. 1970a. *Classification and Biology*. Atherton Press, New York.
Crowson, R. A. 1970b. Review of "Die Stammesgeschichte der Insekten" by W. Hennig. *Syst. Zool.* 19:393–96.
Darlington, P. J. 1970. A practical criticism of Hennig-Brundin "Phylogenetic Systematics" and Antarctic biogeography. *Syst. Zool.* 19:1–18.
Darlington, P. J. 1972. What is cladism? *Syst. Zool.* 21:128–29.
Eldredge, N. 1972. Systematics and evolution of *Phacops rana* (Green, 1932) and *Phacops iowensis* Delo, 1935 (Trilobita) from the Middle Devonian of North America. *Bull. Amer. Mus. Nat. Hist.* 147:45–114.
Eldredge, N. 1973. Systematics of Lower and Middle Devonian species of the trilobite *Phacops* Emmrich in North America. *Bull. Amer. Mus. Nat. Hist.* 151:285–338.
Eldredge, N., and Gould, S. J. 1972. Punctuated equilibria: An alternative to phyletic gradualism. In T. J. M. Schopf (ed.), *Models in Paleobiology*, pp. 82–115. Freeman, Cooper & Co., San Francisco.
Eldredge, N., and Tattersall, I. 1975. Evolutionary models, phylogenetic reconstruction, and another look at hominid phylogeny. In F. Szalay (ed.) *Approaches to Primate Paleontology. Contr. Primat.* 5:218–42.
Engelmann, G. F., and Wiley, E. O. 1977. The place of ancestor–descendant relationships in phylogeny. *Syst. Zool.* 26:1–11.
Gaffney, E. S. 1975. A phylogeny and classification of the higher categories of turtles. *Bull. Amer. Mus. Nat. Hist.* 155:387–436.
Gaffney, E. S. 1977. The side-necked turtle family Chelidae: A theory of relationships using shared derived characters. *Amer. Mus. Novit.* no. 2620, pp. 1–28.
Gaffney, E. S. 1979. Tetrapod monophyly: A phylogenetic analysis. *Carnegie Mus. Bull.*, no. 13, in press.
Ghiselin, M. T. 1975. A radical solution to the species problem. *Syst. Zool.* 27:536–44.
Goodrich, E. S. 1916. On the classification of the Reptilia. *Proc. Roy. Soc. London* 89:261–76.
Goodrich, E. S. 1930. *Studies on the Structure and Development of Vertebrates*. Macmillan Co., London.
Gregory, W. K. 1915. Present status of the problem of the origin of the tetrapods, with special reference to the skull and paired limbs. *Ann. N.Y. Acad. Sci.* 26:317–83.
Gregory, W. K. 1949. The humerus from fish to man. *Amer. Mus. Novit.* no. 1400, pp. 1–54.
Griffiths, G. C. D. 1972. *The Phylogenetic Classification of Diptera Cyclorrhapha, with Special Reference to the Structure of the Male Postabdomen*. W. Junk, The Hague.
Griffiths, G. C. D. 1974. On the foundations of biological systematics. *Acta Biotheor.* 23:85–131.
Harper, C. W. 1976. Phylogenetic inference in paleontology. *J. Paleontol.* 50:180–93.

Hecht, M. K., and Edwards, J. L. 1977. The methodology of phylogenetic inference above the species level. In M. K. Hecht, P. C. Goody, and B. M. Hecht (eds.), *Major Patterns in Vertebrate Evolution,* pp. 3–51. NATO Advanced Study Institutes Series. Plenum Press, New York.

Hempel, C. G. 1966. *Philosophy of Natural Science.* Prentice-Hall, Englewood Cliffs, N.J.

Hennig, W. 1950. *Grundzüge einer Theorie der phylogenetischen Systematik.* Deutscher Zentralverlag, Berlin.

Hennig, W. 1965. Phylogenetic systematics. *Ann. Rev. Entomol.* 10:97–116.

Hennig, W. 1966. *Phylogenetic Systematics.* University of Illinois Press, Urbana.

Hennig, W. 1969. Die Stammesgeschichte der Insekten. Waldemar Kramer, Frankfurt/Main.

Hennig, W. 1975. "Cladistic analysis or cladistic classification?"—A reply to Ernst Mayr. *Syst. Zool.* 24:244–56.

Hull, D. 1970. Contemporary systematic philosophies. *Ann. Rev. Ecol. Syst.* 1:19–54.

Hull, D. 1974. *Philosophy of Biological Science.* Prentice-Hall, Englewood Cliffs, N.J.

Hull, D. 1975. Reviews of "Philosophy of Natural Science" by Carl Hempel, "The Structure of Scientific Revolutions," 2nd. ed., by Thomas S. Kuhn, and "Galileo: A Philosophical Study" by Dudley Shapere. *Syst. Zool.* 24:394–401.

Huxley, J. S. 1958. Evolutionary processes and taxonomy, with special reference to grades. *Uppsala Univ. Arsskrift* 6:21–39.

Jarvik, E. 1952. On the fish-like tail in the ichthyostegid stegocephalians with descriptions of a new stegocephalian and a new crossopterygian from the upper Devonian of East Greenland. *Medd. Groenl.* 114:1–90.

Jarvik, E. 1955. Ichthyostegalia. In J. Piveteau (ed.), *Traité de Paléontologie, Amphibiens, reptiles, oiseaux* 5:53–66. Mason et Cie, Paris.

Kuhn, T. S. 1970. *The Structure of Scientific Revolutions,* 2nd ed. (International Encyclopedia of Unified Science, Vol. 2). University of Chicago Press, Chicago.

Lakatos, I., and Musgrave, A., eds. 1970. *Criticism and the Growth of Knowledge.* Cambridge University Press.

Løvtrup, S. 1977. *The Phylogeny of Vertebrata.* John Wiley and Sons, New York.

MacFadden, B. J. 1976. Cladistic analysis of primitive equids, with notes on other perissodactyls. *Syst. Zool.* 25:1–14.

McKenna, M. C. 1975. Toward a phylogenetic classification of the Mammalia. In W. P. Luckett and F. S. Szalay (eds.), *Phylogeny of the Primates,* pp. 21–46. Plenum Press, New York.

Magee, B. 1973. *Karl Popper.* Viking Press, New York.

Magee, B. 1974. Karl Popper: The useful philosopher. *Horizon* 16:52–57.

Mayr, E. 1969. *Principles of Systematic Zoology.* McGraw-Hill, New York.

Mayr, E. 1974. Cladistic analysis or cladistic classification? *Z. Zool. Syst. Evolutionsforsch.* 12:94–128.

Miles, R. S. 1973. Relationships of acanthodians. In P. H. Greenwood, R. S. Miles, and C. Patterson (eds.), *Interrelationships of Fishes*, pp. 63–103. Academic Press, London.

Miles, R. S. 1975. The relationships of the Dipnoi. In *Problèmes actuels de Paléontologie (Evolution des Vertébrés)*. Coll. Internat. C.N.R.S., no. 218, pp. 133–48.

Moy-Thomas, J. A., and Miles, R. S. 1971. *Palaeozoic Fishes*. Chapman and Hall, London.

Nelson, G. J. 1968. Gill arches of teleostean fishes of the division Osteoglossomorpha. *J. Linn. Soc. Zool.* 47:261–77.

Nelson, G. J. 1969. Gill arches and the phylogeny of fishes, with notes on the classification of vertebrates. *Bull. Amer. Mus. Nat. Hist.* 141:475–552.

Nelson, G. J. 1970. Outline of a theory of comparative biology. *Syst. Zool.* 19:373–84.

Nelson, G. J. 1971a. "Cladism" as a philosophy of classification. *Syst. Zool.* 20:373–76.

Nelson, G. J. 1971b. Review of "Classification and Biology," by R. A. Crowson. *Evolution* 25:441–42.

Nelson, G. J. 1972a. Phylogenetic relationships and classification. *Syst. Zool.* 21:227–31.

Nelson, G. J. 1972b. Comments on Hennig's "phylogenetic systematics" and its influence on ichthyology. *Syst. Zool.* 21:364–74.

Nelson, G. J. 1973a. The higher-level phylogeny of vertebrates. *Syst. Zool.* 22:87–91.

Nelson, G. J. 1973b. Classification as an expression of phylogenetic relationships. *Syst. Zool.* 22:344–59.

Nelson, G. J. 1973c. Relationships of clupeomorphs, with remarks on the structure of the lower jaw in fishes. In P. H. Greenwood, R. S. Miles, and C. Patterson (eds.), *Interrelationships of Fishes*, pp. 333–50. Academic Press, London.

Nelson, G. J. 1974. Darwin–Hennig classification: A reply to Ernst Mayr. *Syst. Zool.* 23:452–58.

Panchen, A. L. 1967. The nostrils of choanate fishes and early tetrapods. *Biol. Rev.* 42:374–420.

Parrington, F. R. 1958. The problem of the classification of reptiles. *Jour. Linn. Soc. Zool.* 44:99–115.

Patterson, C. 1978. Verifiability in Systematics. *Syst. Zool.* 27:218–22.

Platnick, N. I. 1977a. Parallelism in phylogeny reconstruction. *Syst. Zool.* 26:93–96.

Platnick, N. I. 1977b. Review of "Concepts of Species. *Syst. Zool.* 26:96–98.

Platnick, N. I. 1977c. The hypochiloid spiders: A cladistic analysis with notes on the Atypoidea (Arachnida, Araneae). *Amer. Mus. Novit.* no. 2627, pp. 1–23.

Platnick, N. I. 1977d. Paraphyletic and polyphyletic groups. *Syst. Zool.* 26:195–200.

Platnick, N. I. 1977e. Review of "Evolution and the Diversity of Life: Selected Essays" by Ernst Mayr. *Syst. Zool.* 26:224–28.

Platnick, N. I. 1977f. Cladograms, phylogenetic trees, and hypothesis testing. *Syst. Zool.* 26:438–42.
Platnick, N. I., and Cameron, H. D. 1977. Cladistic methods in textual, linguistic, and phylogenetic analysis. *Syst. Zool.* 26:380–85.
Platnick, N. I., and Gaffney, E. S. 1977. Systematics: A Popperian perspective (Reviews of "The Logic of Scientific Discovery" and "Conjectures and Refutations" by Karl R. Popper). *Syst. Zool.* 26:360–65.
Platnick, N. I., and Gaffney, E. S. 1978a. Evolutionary Biology: A Popperian perspective. *Syst. Zool.* 27:137–141.
Platnick, N. I., and Gaffney, E. S. 1978b. Systematics and the Popperian Paradigm. *Syst. Zool.* 27:381–88.
Platnick, N. I., and Gertsch, W. J. 1976. The suborders of spiders: A cladistic analysis (Arachnida, Araneae). *Amer. Mus. Novit.* no. 2607, pp. 1–15.
Popper, K. R. 1960. *The Poverty of Historicism.* Routledge and Kegan Paul, London.
Popper, K. R. 1968a. *The Logic of Scientific Discovery,* 2nd ed, Harper and Row, New York.
Popper, K. R. 1968b. *Conjectures and Refutations.* Harper and Row, New York.
Popper, K. R. 1972. *Objective Knowledge, an Evolutionary Approach.* Clarendon Press, Oxford.
Popper, K. R. 1976. *Unended Quest: An Intellectual Autobiography,* rev. ed. Open Court Publishing Co., La Salle, Illinois.
Rensch, B. 1959. *Evolution above the Species Level.* Columbia University Press, New York.
Romer, A. S. 1947. Review of the Labyrinthodontia. *Bull. Mus. Comp. Zool.* 99:1–368.
Romer, A. S. 1956. *Osteology of the Reptiles.* University of Chicago Press, Chicago.
Rosen, D. E. 1973. Interrelationships of higher euteleostean fishes. In P. H. Greenwood, R. S. Miles, and C. Patterson (eds.), *Interrelationships of fishes.* Academic Press, London.
Rosen, D. E. 1974a. Review of "Classification and Biology," by R. A. Crowson. *Syst. Zool.* 23:141–43.
Rosen, D. E. 1974b. Cladism or gradism?—A reply to Ernst Mayr. *Syst. Zool.* 23:446–51.
Rosen, D. E. 1974c. Phylogeny and zoogeography of salmoniform fishes and relationships of *Lepidogalaxias salamandroides. Bull. Amer. Mus. Nat. Hist.* 153:267–323.
Säve-Söderbergh, G. 1934. Some points of view concerning the evolution of the vertebrates and the classification of this group. *Ark. Zool.* 26:1–20.
Schaeffer, B. 1965. The rhipidistian–amphibian transition. *Amer. Zool.* 5:267–76.
Schaeffer, B., Hecht, M. K., and Eldredge, N. 1972. Paleontology and phylogeny. *Evol. Biol.* 6:31–46.
Schilpp, P. A., ed. 1974. *The Philosophy of Karl Popper.* Open Court, La Salle, Ill.
Simpson, G. G. 1975. Recent advances in methods of phylogenetic inference. In W. P. Luckett and F. S. Szalay (eds.), *Phylogeny of the Primates,* pp. 3–19. Plenum Press, New York.

Sokal, R. R. 1975. Mayr on cladism—and his critics. *Syst. Zool.* 24:257–62.
Suppe, F., ed. 1977. *The Structure of Scientific Theories*, 2nd ed. University of Illinois Press, Urbana.
Szalay, F. S. 1977. Ancestors, descendants, sister groups and testing of phylogenetic hypotheses. *Syst. Zool.* 26:12–18.
Tattersall, I., and Eldredge, N. 1977. Fact, theory, and fantasy in human paleontology. *Amer. Sci.* 65:204–11.
Thompson, K. S. 1966. The evolution of the tetrapod middle ear in the rhipidistian–amphibian transition. *Amer. Zool.* 6:379–97.
Thomson, K. S. 1967. Mechanisms of intracranial kinetics in fossil rhipidistian fishes (Crossopterygii) and their relatives. *J. Linn. Soc. Zool.* 46:223–53.
Westoll, T. S. 1943. The origin of tetrapods. *Biol. Rev.* 18:78–98.
Wilder, B. G. 1887. Remarks on classification of Vertebrata. *Amer. Nat.* 21:913–17.
Wiley, E. O. 1975. Karl R. Popper, systematics, and classification: A reply to Walter Bock and other evolutionary taxonomists. *Syst. Zool.* 24:233–42.
Wiley, E. O. 1976. The phylogeny and biogeography of fossil and recent gars (Actinopterygii: Lepisosteidae). *Misc. Publ. Univ. Kan.*, no. 64, pp. 1–111.

# Recognition of Ancestor–Descendant Relationships in Invertebrate Paleontology

## Sara S. Bretsky

Cracraft (1974) has claimed that two competing models of phylogeny reconstruction are in use today. He perceives three major areas of disagreement among practitioners of the two schools:

1. On the question of classification—whether relative recency of common ancestry should be the sole criterion for defining or ranking taxa, or degree of morphological divergence should also be considered in setting up classifications;
2. On the question of how species have originated, whether predominantly by allopatric speciation or gradual *in situ* transformation of one species into another;
3. On the question of recognition of ancestor–descendant relationships— whether actual fossil organisms should be identified as the ancestors of other fossil or of living species; and if this is permissible in principle, how can ancestor–descendant relationships be recognized in practice?

Of these three points of apparent conflict, the differences in approach to classification have previously been thoroughly discussed (Bock, 1973, Nelson, 1973) and require only cursory mention here. Briefly, phylogenetic systematists advocate treating all taxa as though they were contemporaneous and recognizing relationships among taxa only on the basis of relative recency of common ancestry, which is to be inferred from the distribution of shared derived states of presumably homologous characters (Schaeffer, Hecht, and Eldredge, 1972). Similarity based on shared possession of ancestral character states is not to be considered as a classificatory criterion (Cracraft, 1974, Eldredge and Tattersall, 1975; Hennig, 1966). (For example, if the hypothetical ancestral mammal is inferred to have had five digits, then possession of five digits, as a shared ancestral character state, is not a legitimate basis for defining groups within the Mammalia.)

Evolutionary systematists, by contrast, advocate grouping species and higher taxa on the basis of phenotypic similarity, which is considered to reflect underlying genetic similarity (Bock, 1973, p. 377; Mayr, 1965, pp. 79, 89) and also to bear at least a rough correspondence to relative recency of common ancestry (Simpson, 1961, pp. 113, 141). Organisms having diverged greatly in morphology from an ancestral taxon, however, may be given a higher categorical rank than less divergent descendants of the common ancestor (Mayr, 1969, pp. 70, 231; Simpson, 1961, p. 195). In contrast to a third group of systematists—the pheneticists or numerical taxonomists (Sokal and Sneath, 1963; Sneath and Sokal, 1973)—evolutionary systematists also stress the importance of subtracting from "overall" similarity its convergent component, i.e., morphological resemblance resulting from adaptation to the same habitat or similar habitats rather than from inheritance from a common ancestor.

The second source of apparent disagreement concerns the predominant mode of origin of species. But difference of opinion on the question of "punctuated equilibria" versus "phyletic gradualism" (Eldredge and Gould, 1972; Gould and Eldredge, 1977), or, more precisely, on the relative importance of the two processes, has some deficiencies as a diagnostic character for separating phylogenetic systematists from evolutionary systematists. The founding fathers of phylogenetic systematics have expressed ideas on speciation as gradualistic as those of any paleontologist (cf. Crowson, 1970, pp. 42–43; Hennig, 1966, pp. 58–60, figs. 14 and 15). In appropriating the model of punctuated equilibria for phylogenetic systematics, however, Cracraft (1974, p. 74) argued that, since examining individual lineages of fossils for evidence of phyletic gradualism involves the recognition of ancestors, phyletic gradualism is inextricably mingled with the evolutionary systematic model. Yet one of the coauthors of the paper which first distinguished "punctuated equilibria" from "phyletic gradualism" is an unabashed evolutionary systematist (Gould, 1973, p. 324; Gould, 1975b, pp. 244–45).

Thus accepting phyletic gradualism has, historically, not implied embracing evolutionary systematics or vice versa. And, in subsequent sections of this paper, I shall argue that the phylogenetic systematic model, by employing an unrealistically simple portrayal of character change in the process of speciation, sets up a false dichotomy between inferring genealogical continuity among conspecific individuals (perceived as very easy) and inferring similar continuity among individuals of different species (perceived as impossible). I shall attempt to demonstrate that ancestor–descendant relationships among taxa can be inferred even if one accepts an episodic (punctuational) view of speciation events, because arguments invoked to document such relationships do not differ in principle from those necessary to document conspecificity among individuals belonging to different morphs of a single species.

Discussion of Cracraft's third distinction between evolutionary and phylogenetic systematists, the question of whether or not to recognize—and if so, how to recognize—ancestor–descendant relationships from evidence in the fossil record will thus predominate in subsequent sections of this paper. Tacitly, the organization of this symposium suggests that invertebrate paleontologists and vertebrate paleontologists, even if they adhere to the same theoretical model, may envision the process of phylogenetic reconstruction differently. In attempting to summarize the perspective of invertebrate paleontologists toward problems of phylogeny reconstruction, I have concluded that, for many invertebrate paleontologists, phylogeny reconstruction has not been recognized as being a theoretical problem.

The traditional position of invertebrate paleontology as the "handmaiden of stratigraphy" implicitly fosters a simple, straightforward notion of the relationship between stratigraphic position and phylogenetic evidence, namely, that those species occurring in older beds must be directly ancestral to morphologically similar ones occurring in younger strata of the same geographic region. A quotation from one of the classic works in evolutionary paleontology (Stenzel, 1949, p. 34) illustrates this way of looking at things: "Only recently has it become possible to trace this succession of species [of Eocene oysters], because only recently has the stratigraphic succession of the beds and localities involved been cleared up sufficiently to allow correct chronologic placing of the species involved." In other words, given adequate chronological documentation, the ancestor–descendant relationships between species fall into place almost automatically. Phylogenetic inference required only describing stratigraphically defined evolutionary trends and, if possible, accounting for their nature and direction (Why the increase in size? Why fewer plicae on the shell? and so on).

Summaries such as those of Cracraft (1974), which contrast an "evolutionary systematic" and a "phylogenetic systematic" model and implicitly assign all paleontologists who are not avowed "pheneticists" or "cladists" to the "evolutionary systematic" category, may involve an oversimplification analogous to that of the customary formula for placing infants with adoption agencies (assign the child to the agency representing the parents' religious affiliation; if the relevant information is unknown, the child is a Protestant). By regarding "conventional invertebrate paleontology" and "evolutionary systematics" as synonymous, one may similarly be overestimating the degree of actual active commitment to a particular theoretical model.

When I speak of conventional or common practice in invertebrate paleontology, therefore, I shall sometimes be referring to research in invertebrate paleontology as it has been influenced by the ideas and principles articulated most fully by Simpson (1961) and Mayr (1963, 1969). But at other times, I shall be speaking of a prior and still essentially independent tradition, with few theoretical underpinnings, except for a faith that collecting more and more fossils and correlating more and more local sections ultimately leads to an accurate depiction of the course of evolution (Levinton, 1976, provides comments on this attitude). One cannot really describe the evolutionary systematic approach to phylogeny reconstruction in invertebrate paleontology without acknowledging that much theoretical work in invertebrate paleontology has gone on with, figuratively speaking, one ear cocked for reactions from neontologists and vertebrate paleontologists and the other for responses from practical biostratigraphers. One may view the first group of workers as having strong theoretical backgrounds but as usually lacking access to data amenable to testing hypotheses about species-level change in evolutionary time, and the second group as possessing much good information which has seldom been collected or analyzed in the context of ecological and evolutionary models.

An example of the dichotomy occurs precisely on the important question of how well the stratigraphic sequence of character states ("chronocline") reflects the true "spectrum of primitive–derived character states" (Schaeffer, Hecht, and Eldredge, 1972, p. 33). Invertebrate paleontologists presumably have long been aware (cf. Bretsky, 1975, p. 113; Colbert, 1961, p. 108) that, in principle, stratigraphic position may be a deceptive guide to phylogenetic relationships. But many of us whose interests lie preeminently in systematics and evolution may have consciously chosen to work with marine invertebrate fossils precisely in the hope of avoiding some apparent obstacles (small sample size, sporadic preservation often under unusual environmental conditions, poor representation of terrestrial facies in the geologic record, difficulties in making interregional correlations) to examining species-level details of the evolution of vertebrates or of poorly fossilizable invertebrates. Meticulous stratigraphic analysis promised the reward of "a continuum of morphologies between . . . taxa with preservable parts" (Durham, 1967, p. 559). One's choice of organisms, times, and places for phylogenetic investigation was often dictated or at least strongly influenced by how closely one expected one's collections to approach such a continuum.

Perhaps this research strategy grew out of an undue complacency about the connection between stratigraphic position and phylogenetic significance. Yet expectations have colored perceptions (Eldredge and Gould, 1972, p. 83) in both phylogenetic systematics and more-traditional biostratigraphic paleontology. Phylogenetic systematists see the course of change in character states as generally forming a unique unidirectional sequence (morphocline of Maslin, 1952) from ancestral to most highly derived (fig. 7; Cracraft, 1974, fig. 2; Eldredge, 1972, fig. 17; Hennig, 1966, fig. 21). If the stratigraphic sequence of characters of fossil taxa disagrees with inferred morphocline polarity, so much the worse for the fossil record. In contrast, invertebrate paleontologists have accepted the order of appearance of fossil taxa as a fairly literal reading of phylogenetic history. Under this premise, and perhaps also to emphasize their rejection of the orthogenetic speculations so thoroughly discredited by Simpson (1949, pp. 130–59), evolutionary systematic workers in invertebrate paleontology have envisioned frequent occurrence of evolutionary reversal in individual character states (e.g., Bretsky, 1976, fig. 13), of parallelism (e.g., Newell and Boyd, 1975, fig. 1), and of iterative evolution (e.g., Gould, 1969, fig. 20). If assumptions about the direction of change in character states conflicts with the observed temporal order of taxa, so much the worse for morphocline polarity.

These contrasting expectations make it difficult to establish a dialogue, much less a reconciliation, between the insights of a clearly defined, internally consistent, phylogenetic systematic *model* and the practice of a much more diffuse *profession* of what one might call, following Kuhn (1962), "normal" invertebrate paleontology. (I prefer this adjective to "conventional" or "traditional," since much work on, for example, paleoecological reconstruction, stratigraphic correlation, analysis of morphological function, or comprehension of broad-scale patterns of diversification and extinction has been distinctly nontraditional and innovative but has only peripherally concerned systematics and phylogeny). Nevertheless, dialogue may perhaps be initiated by considering four points which phylogenetic systematists appear to regard as insuperable difficulties in using the geologic record as a guide to ancestor–descendant relationships. I schematically identify these as (1) incomplete representation of species diversity, (2) inaccurate reflection of species range, (3) disjunct geographic distribution of ancestral and descendant species, and (4) morphological discontinuity between ancestor and descendant. A fifth issue, not necessarily involving disagreement about the interpretation of stratigraphic evidence, but growing logically out of the investigation of morphological variability within and among taxa (point 4 above), is the question of recognizing ancestor–descendant relationships among supraspecific taxa. I shall discuss each of these five issues in turn, attempting to evaluate both the theoretical basis for them and the problems which they may pose in practice. My aim is less to expose as illusory the objections to use of stratigraphic evidence than to assess how serious such objections are in specific paleontological applications. I shall illustrate such applications principally through examples drawn from my own work on Tertiary bivalve mollusks or from work of others when I feel reasonably competent to evaluate potential taxonomic or stratigraphic difficulties. No complete review of the literature on ancestor–descendant relationships among invertebrate fossils is contemplated or intended.

## Incomplete Representation of Species Diversity

Engelmann and Wiley (1977) point out that if the relative temporal sequence of two sister species (fig. 1a) is accurately known, there are two possible relationships between the older species A and the younger species B. Either A is directly ancestral to B (fig. 1b) or A and B are both descended from a common ancestral species not represented in the fossil sample studied

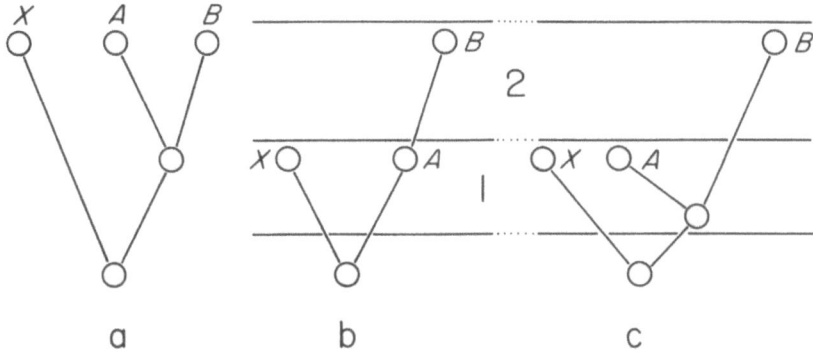

Figure 1. Alternative interpretations of a relationship between two sister species, A and B, when the stratigraphic sequence of their occurrence is known. (a) Cladogram indicating that A and B are more closely related to each other than to other known species of their taxon, collectively symbolized by X. (b) Species A (occurring in stratigraphic horizon 1) interpreted as ancestral to B (occurring in stratigraphic horizon 2). (c) Species A and B interpreted as descended from an unknown common ancestral species. (After Engelmann and Wiley, 1977.)

(fig. 1c). Schaeffer, Hecht, and Eldredge (1972) have similarly illustrated the problems that incomplete representation of species diversity poses in phylogenetic reconstruction. Their diagram (reproduced as fig. 2) shows that, even if the correct stratigraphic order of fossil species (branches marked by X's in the diagram) is known, these species need not bear a direct genealogical relationship to each other. They may, rather, have been connected by intermediate species which have left no fossil record.

Formulations such as these two prompt the question, How imperfect *is* the fossil record? How likely is it that an ancestral species will totally fail to be represented in a given fossil sample? Obviously, the likelihood of this happening varies enormously from taxon to taxon, and from time to time in geologic history. Evaluating bias in the geologic record of species diversity has been much discussed (Bambach, 1977; Kier, 1977; Raup, 1972, 1976; Valentine, 1970). The topic has seldom been explored, however, in the context of studies of actual lineages which would correspond to the hypothetical lineage in figure 2.

Durham (1967) estimated that for marine invertebrates with hard parts (presumably the kind of organisms most likely to be fossilized), somewhere between 1 in 44 and 1 in 100 of all species which have ever lived are actually known from fossils. Within a single higher taxon from evidently comparable

Figure 2. A hypothetical phylogenetic tree. The vertical axis is time, the horizontal axis morphological differentiation. Species represented by known fossil samples are indicated by branches marked X; blank branches correspond to species which have left no fossil record. Note that, although they may be arranged in correct temporal sequence, species known from fossils may not be directly ancestral to other such species. (After Schaeffer, Hecht, and Eldredge, 1972.)

Table 1. Number of Recent Molluscan Species in the Panamic Province and Fossil Molluscan Species in Various Epochs of the California Tertiary[1]

|  | Paleocene | Eocene | Oligocene | Miocene | Pliocene | Recent |
|---|---|---|---|---|---|---|
| Number of species | 170 | 630 | 260 | 370 | 300 | 2,500 |
| Percentage of Recent figure | 6.8 | 25.2 | 10.4 | 14.8 | 12.0 |  |

[1] Data from Durham, 1967.

climatic regimes, he found the apparent bias to be less severe but still substantial. Comparing the number of molluscan species in the Recent Panamic province to the number described for each of the Tertiary epochs in California, he showed that species diversity for these time periods was only 1/4 to 1/16 of that of the Recent moment (table 1).

I have carried out an analogous, but more narrowly focused, evaluation of bias in the representation of species diversity, concentrating not only on a single basin of deposition (comprising approximately the same area, range of lithologies, and climatic conditions throughout the time period covered) but also on a single lineage of organisms. The data here presented (table 2) were originally compiled in answer to a question about turnover rate of fossil species. The taxon used is a family of bivalve mollusks, the Lucinidae. I compare the number of living lucinid species in the northern West Atlantic to the number of lucinid species recorded for each of the Tertiary epochs from strata of the Atlantic and Gulf Coastal Plain of the United States. Because the placement of the Pliocene–Pleistocene boundary in the Coastal Plain is controversial (Blackwelder, 1972; Oaks and DuBar, 1974), I combine figures for species recorded in the literature as Pliocene and Pleistocene. Kier (1977) adopts a similar expedient in studying the observed diversity of regular and irregular echinoids for Mesozoic and Cenozoic time.

Species assignments for the Recent lucinids come from a study by Britton (1970), and I concur with him on all essential questions of species-level taxonomy. I omit two species whose sole reports in the literature Britton regards as based on beachworn specimens of fossil species, as well as two species discussed by him which are known only from Brazil. (Several of the Recent northern West Atlantic species do range this far south.) Britton (1970, table 1) fully documents recorded depth ranges for the Recent lucin-

Table 2. Subgenera and Their Constituent Species, by Epoch, of Lucinidae from Tertiary Strata of the Atlantic Costal Plain and from the Recent Northern West Atlantic [1]

| | Paleocene (13) | Eocene (21) | Oligocene (7) | Miocene (33) | Plio-Pleistocene (29) | Recent (27) |
|---|---|---|---|---|---|---|
| *Lucina (Lucina)* | | *carinifera hamata* | *wacissana* | *wacissana glenni* | *pensylvanica densata* | *pensylvanica* |
| *Lucina (Stewartia)* | | | | *anodonta foremani nocariensis* | *anodonta wongi floridana* | *floridana* |
| *Lucina (Callucina)* | *temprana evansae sabelli* | { *papyracea curta ulrichi whitei perminuta* | *choctavensis* | *cala actinoides prunoides* | *radians* | *radians* |
| *Lucina (Parvilucina)* | | | | *crenulata* { *sphaeriola piluliformis vaughani flumenvadosa diktyota chipolana* | *crenulata* { *multilineata multistriata pectinella* | *multilineata pectinella costata* |
| *Lucina (Cavilinga)* | | *pomilia atoma* | | *trisulcata* { *parawhitfeldi pruna recurrens* | *trisulcata blanda* | *blanda* |

| | | | | |
|---|---|---|---|---|
| Lucina (Radiolucina) | | | | amiantus |
| | | | waccamawensis | |
| | | | tuomeyi | |
| | | | amiantus | |
| Lucina (Lucinisca) | | | cribraria | nassula |
| | | | nassula | |
| | | { plesiolopha | | |
| | | calhounensis | | |
| | | chipolana | | |
| Lucina (Pleurolucina) | | | amabilis | leucocyma |
| | | | choctawhatcheensis | sombrerensis |
| | | | | hendersoni[2] |
| Lucina? (Recurvella) | dolabra | | | |
| Phacoides (Phacoides) | juliannae | | pectinatus | pectinatus |
| | fortidentalis | | | |
| Phacoides (Plastomiltha) | claibornites | | | |
| | gaufra | | | |
| Phacoides (Armimiltha) | | hillsboroensis | discformis | |
| | | { hillsboroensis | | |
| | | beilprini | | |
| Phacoides (Lucinoma) | | | contractus | |
| | | contractus | | |
| Phacoides (Thora) | ozarkana | | | |
| Codakia (Codakia) | hanseni | | | filosus |
| | oblonga | | | blakeanus[2] |
| | | | | atlantis[2] |
| Codakia (Ctena) | phaedra | | orbicularis | orbicularis |
| | aricia | | | |
| | | magnolioides | { orbiculata | orbiculata |
| | | scurra | speciosa | |
| | | | magnoliana | |
| | | | microimbricata | |
| | | | leonensis | |

Table 2 (*continued*)

| | Paleocene (13) | Eocene (21) | Oligocene (7) | Miocene (33) | Plio-Pleistocene (29) | Recent (27) |
|---|---|---|---|---|---|---|
| *Codakia?* (*Claibornites*) | | *symmetrica* | | | | |
| *Miltha* (*Miltha*) | | *greggi* | *chipolana perovata* | *chipolana* | *caloosaensis carmenae* | |
| *Miltha* (*Codakincina*) | *aquiana claytonia* | | | | | |
| *Miltha* (*Eomiltha*) | | *pandata megameris ocalana* | | *xustris scolaroi* | | |
| *Myrtea* (*Myrtea*) | *mesakta* | | | | | *sagrinata* [2] *lens* |
| *Myrtea* (*Myrteopsis*) | | | | | | *compresa* [2] |
| *Myrtea* (*Gonimyrtea*) | *ubleri* | *ubleri biscuplta* | *taylorensis vicksburgensis* | *waltonensis* | | *pristiphora* |
| *Myrtea?* (*Gardnerella*) | | | | | | |
| *Anodontia* (*Anodontia*) | | *augustana* | | *janus corpulenta santarosiana* | *alba* | *alba schrammi phenax* |
| *Anodontia* (*Eophysema*) | | *subvexa dartoni* | | | | |
| *Divaricella* (*Divalinga*) | | | | *chipolana waltonia* | *compsa quadrisulcata* | *quadrisulcata* |
| *Divaricella* (*Egracina*) | | | | | | *dentata* |

[1] Numbers below the name of each time period are the total number of known lucinid species for that time period, if the bracketed species are regarded as synonyms.
[2] Recent species known only from water depths greater than 300 ft.

ids; I indicate in table 2 only those species occurring at water depths greater than 300 feet and thus in habitats not likely to be represented in Coastal Plain strata. Distributional data for the Tertiary lucinids are given more fully in Bretsky (1976) and Kauffman and Bretsky (unpublished manuscript).

Comparisons of Recent and fossil diversities are often vitiated by a wide disparity in the taxonomic treatment of fossil and Recent species. Recent species may often be more narrowly defined than fossil ones, either because of more zealous taxonomic splitting by collectors or simply because having soft-part as well as shell data permits finer taxonomic discrimination. I believe this source of potential difficulty to be minimal for the North American Tertiary and Recent lucinids, both because I have worked with collections of the Recent as well as fossil species and because diagnoses of the living species are all based principally on shell characters.

For the fossil species, table 2 indicates synonymization of various nominate species which I believe (though I have not yet verified this through morphometric analysis) to be merely variants of a single polymorphic species. Most of these names come from two or three monographs on the Miocene fauna. I also enumerate only species recorded from the continental United States. The coverage of fossil faunas from the Caribbean Islands and northern South America varies widely in terms of intensity of study as well as quality of stratigraphic documentation. I suspect that many nominate Tertiary species from these areas are synonyms of U.S. Coastal Plain species, but again, I lack morphometric confirmation. For consistency, I also omit records of a particular Coastal Plain species for a particular time period when such a combination occurs only outside the continental United States. This exclusion primarily involves records of a few species which seem not to have been found in the Coastal Plain Pleistocene, but which do occur as fossils on Caribbean islands.

Because of the synonymizations and omissions outlined above, the figures for observed diversity of fossil species are conservative relative to what I would have obtained from a simple literature compilation of species names and associated times of occurrence. Nevertheless, the "raw" diversity data (figures at the top of table 2) show that for three time periods—the Eocene, Miocene, and Plio-Pleistocene—total diversity is comparable in magnitude to Recent standing diversity. The observed Paleocene and Oligocene lows (particularly the latter) may be exaggerated locally by factors such as smaller outcrop area of sediments of these ages, poorer preservation of the faunas, or more limited accessibility of exposures. Nevertheless, many other tabulations of observed species diversity for different sorts of organisms and different geographic areas, as well as for the Coastal Plain Tertiary (e.g., Cifelli, 1969; Durham, 1967; Kier, 1977), also show lower diversities in the Paleocene and Oligocene. Fischer and his colleagues (Fischer, 1977; Thompson, 1977) find evidence for a Late Cretaceous–Early Tertiary and a mid-Tertiary decrease in diversity for many different phyla so convincing as to fall into a suggested cyclicity of faunal crises, approximately 32 million years in period. Whether or not one accepts the argument for cyclicity, the observed decreases in number of species for many different sorts of organisms do cover a broad enough spectrum of facies, quality of preservation, and sampling localities not to be dismissed as artifactual. Thus our particular instance, the smaller number of lucinid species recorded for the Coastal Plain Paleocene and Oligocene, probably does result from actual lower diversities at these times rather than being entirely a consequence of biased sampling.

The detailed breakdown of occurrence of species within subgenera for each epoch shows only a few gaps between first and last appearance in Atlantic Coastal Plain strata (or between first appearance and the Recent) for the 28 lucinid subgenera of table 2, given that the temporal sampling intervals are defined as Tertiary epochs. This look at the pattern of both the "raw" diversity tabulation and the components of these figures suggests two conclusions:

1. Contrary to the implications of Durham's estimate of how incomplete our knowledge of the fossil record is, we are not dealing with order-of-magnitude discrepancies between the number of known fossil and Recent lucinid species. If the percentage discrepancies between number of fossil and Recent species which Durham observed for the West Coast Cenozoic mollusks had held good for the lucinids, we would have expected to see, for example, only 1 or 2 Paleocene lucinid species (we actually have 13), 5 or 6 Eocene species (in contrast to 21), and so on.

2. By looking at the detailed breakdown of species diversity through time within probably monophyletic groups, we can see which of these higher taxa approach a "dense and continuous" (Gingerich, 1977) fossil record at the species level, and for which ones there are substantial gaps in the fossil record. I have previously argued (Bretsky, 1971, p. 217, 1976, p. 232) that for most of the Coastal Plain occurrences, the precision of stratigraphic resolution is as good as or better than the expected duration

of species. When this condition is fulfilled, paleontologists have assumed that morphologically similar species from adjacent time periods can be combined into lineages. My strategy for working out lucinid phylogeny thus has corresponded to the "stratophenetic" approach of Gingerich (1976, 1977), though at the level of species rather than of stratigraphically successive samples within species.

Clearly, simply finding an approximate correspondence between diversity of fossil species throughout a given span of time and diversity of comparable Recent species (as with the Eocene, Late Tertiary, and Recent lucinids of table 2) does not demonstrate that diversity has remained constant through time. For a group such as the Cenozoic bivalves, though, there seems to be no geological or ecological reason to expect massive disappearances of species that have left no trace in the fossil record.

Even if one accepts Recent standing diversity as an adequate predictor of the expected level of past diversity of an extant group during a circumscribed span of time, for taxa which are either extinct or unquestionably much less common today there is no obvious "baseline" to use in such a comparison. To what shall we compare a family of trilobites or of Paleozoic nautiloids? Granting for the sake of argument that one could make a convincing case for regarding a particular extant taxon as a satisfactory morphological or ecological analog of an extinct one (e.g., teleost fishes and orthoconic cephalopods, isopods and trilobites), the species diversity of the chosen Recent taxon could be substantially higher than that of its putative fossil analog at the corresponding level of the taxonomic hierarchy.

Bambach (1977) estimates that the total number of invertebrate species in the Recent marine realm is about four times that in the Lower Paleozoic. He also presents convincing evidence that the apparent increase in marine invertebrate diversity has occurred in a stepwise fashion rather than at a constant rate and that it has affected offshore (presumably lower-stress) environments more than nearshore (presumably higher-stress) ones. Thompson (1977, p. 29) similarly proposes that, despite the evident oscillations in diversity of the marine realm throughout the Mesozoic and Cenozoic (such as the Paleocene and Oligocene diversity lows of table 2), "oceanic ecosystems are continuing to diversify as new groups are added . . . that divide resources more finely and more efficiently than the groups they replace."

If these patterns of overall increase in diversity throughout the Phanerozoic are real, one does not know whether they prevail within a particular level of the taxonomic hierarchy as well as among levels. If things have changed in an ecological and evolutionary sense[e.g., a greater level of provinciality (Valentine, 1970) or the evolution of more efficient mechanisms for resource utilization (Thompson, 1977)], so that the apparent across-the-board increase in Phanerozoic diversity prevails within particular groups as well, the Recent standing diversity of, say, a family of deep-sea isopods might be several times that of a family of late Cambrian trilobites. [The comparison used here is based on suggestions of an ecological and morphological analogy between isopods and trilobites (Cook and Taylor, 1975; Menzies, George, and Rowe, 1973; Taylor, 1977).] A compilation and comparison of diversity within familes for organisms this widely separated in time might indicate a bias in the fossil data which was either nonexistent or of considerably lesser magnitude than the "raw" diversity comparison would suggest. Conversely, the Recent taxon might be more finely split than a fossil taxon of comparable morphological complexity (Schopf et al., 1975) because of study by more workers, the possibility of basing discriminations on soft as well as hard parts, and all the other factors which make comparing fossil and Recent diversities tricky (Bambach, 1977, p. 155).

Despite all these caveats, an explicit presentation like that in table 2 of the temporal distribution of species within a lineage should at least have heuristic value in pointing up obvious gaps in the species-level record of the taxa composing that lineage. As Valentine (1977, p. 324) points out, if, say, as many as 5 percent of all the marine invertebrate species with hard parts which have ever lived are preserved as fossils and if the occurrences of these are distributed randomly through Phanerozoic time, then only about 0.25 percent of all possible pairs of species are expected to represent ancestor–descendant species pairs. But occurrences of fossil species are not random in time or in space—the quality of the fossil record varies considerably not only (as is well known) among taxa but also within taxa, even at a fairly low level of the taxonomic hierarchy.

Lucinid bivalves, for example, originate in the latter part of the Ordovician and are reasonably common in Devonian strata. But for the whole span of time (some 150 million years) between the end of the Devonian and the beginning of the Jurassic, only about 10 to 15 species of Lucinidae (mostly so poorly preserved as to be identifiable only as to family) have been described. Did lucinid species diversity fall so low during this time span as to be practically indistinguishable from zero? Or was diversity always approximately as high as now, with almost all the specimens representing Carboniferous through Triassic species buried in the subsurface or vanished into subduction zones? One doesn't know, but it is obvious that a stratophenetic approach cannot be used to investigate details of lucinid phylogeny during the Late Paleozoic and Triassic, since the gaps between recorded species occurrences are so much greater than the expected duration of species ranges—the topic to which we now turn.

## Inaccurate Reflection of Species Ranges

Incomplete representation of species diversity is one aspect of "imperfections of the geological record"; the other side of the coin is the question of inaccurate reflection of species ranges. Figure 3 (modified from Harper,

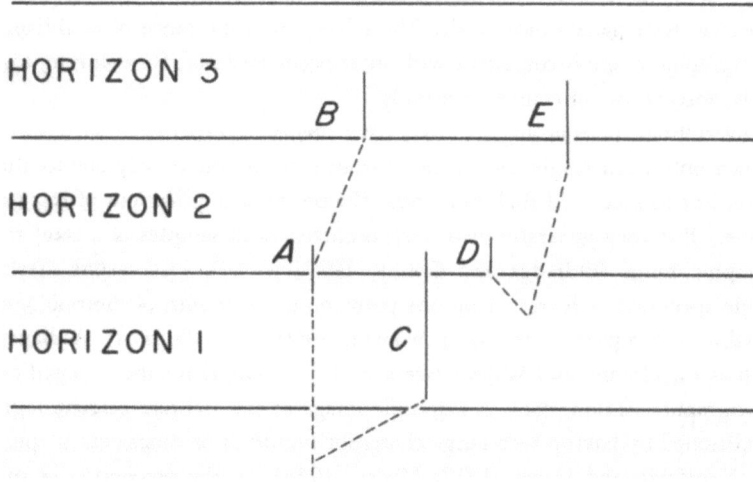

Figure 3. Two hypothetical lineages. Solid lines symbolize the recorded ranges of fossil species, dashed lines the portion of a species' total temporal duration not represented by known fossils. Vertical axis is time, horizontal axis is morphological differentiation. Species A (persisting with little change) has given rise to species C and B. Species D and E have both diverged somewhat from their common ancestral population. Note that, although A actually originated before C, the stratigraphic record of C is more complete than that of A; stratigraphic sequence thus suggests the reverse of the true ancestor–descendant relationship between the two species. (After Harper, 1976.)

1976) diagrams this latter problem. In the hypothetical lineages of figure 3, all the species involved are known from fossils, but the ranges of the species are not equally well known. Solid lines symbolize the recorded ranges of species, dotted lines that portion of the total temporal duration from origin to extinction which is *not* recorded by fossils. The observed stratigraphic sequence of first appearance of species A and C is the reverse of the order in which they actually originated; their chronological occurrence thus suggests that A descended from C, the reverse of the true ancestor–descendant relationship.

To stratigraphic paleontologists, establishing the temporal ranges of fossil species is the principal *raison d'être* of their life's work. By contrast, and with some exceptions (especially Eldredge, 1971; Eldredge and Gould, 1977), phylogenetic systematists simply do not write as though they view fossil species as having a discernible duration in time. Arguments against using stratigraphic position as a key to ancestor–descendant relationships (fig. 1; Cracraft, 1974, fig. 2) are couched in terms of finds of fossil species at a single place and, literally, a single "point in time." Perhaps this view proceeds from the nature of the paleontological material with which most phylogenetic systematists have dealt. The relative unimportance of establishing stratigraphic range is consistent with most occurrences of, for example, fossil birds, insects, or, of course, hominids.

Invertebrate paleontologists, too, have plenty of experience with species known only from single specimens or from a collection at only one locality. (Later in this section, I shall talk about the practical implications of just such a case.) But they generally view such occurrences as samples of a total stratigraphic range (Eldredge and Gould, 1977, p. 27), just as one views a single specimen as representing one point on a continuum of morphological variability (Simpson, Roe, and Lewontin, 1960, pp. 105, 174). A diagram such as Engelmann and Wiley's (see Fig. 1), a cladogram superimposed on a stratigraphic section, conveys a specific image to invertebrate paleontologists conditioned by having seen range charts for hundreds or thousands of species (cf. Kauffman and Hazel, 1977; Moore, 1948). In the perception of most paleontologists, if the relationship between species A, species B, and the unknown common ancestor is as diagrammed in figure 1c, the total temporal range of species B encompasses both level 1 and level 2. A single occurrence of species B in level 2 does not necessarily immutably define the extent to which species B may be represented by known fossils, since, by ordinary assumptions of stratigraphic paleontology, it is always possible that further sampling will turn up specimens of species B from level 1.

Few phylogenetic systematists in recent years have sounded quite as dubious about paleontology (or paleontologists) as Crowson (1970), most of whose discussion of the classification of fossils deals with famous frauds. But many (e.g., Farris, 1976; Nelson, 1973), do tend to talk about fossils as though their existence required complicated classificatory contortions to fit into the Linnaean scheme, and they sometimes sound as though, if forced to choose, they would opt for chucking out fossils before chucking out hierarchical classification.

A little healthy skepticism about the literalness with which chronological order of occurrence of fossil species actually records the true sequence of origination cannot come amiss in invertebrate paleontology. But critiques of the usefulness of the fossil record have to be couched in terms comprehensible to practitioners of biostratigraphy, and presenting such a critique as though all fossil species' durations were equally well (or equally poorly) documented simply is not realistic. In terms of establishing species' ranges, the characteristic sampling problem in invertebrate paleontology is less commonly the occurrence in only one sedimentary layer of a species whose total lifespan was some hundreds of thousands to several millions of years, and more often the discontinuous distribution of such a species (perhaps because of unfavorable environmental conditions in a geographically circumscribed area). Hazel (1977) provides an apt illustration of such a circumstance, in which microfossil species present in abundantly fossiliferous lower sand and upper limestone units in the American Gulf Coast Oligocene are absent from the intervening sparsely fossiliferous clay. Granted adequate sampling from sediments representing the sorts of habitats usually amenable to good fossil preservation (e.g., reefs or level-bottom shelf settings as opposed to the deep sea or rocky shores), invertebrate paleontologists generally tend to regard the "bases" and "tops" of species occurrences as reasonably reliable. Differences in the observed durations of the fossil records of particular species often seem to reflect real differences in ecology and adaptive strategy (T. Hansen, personal communication, 1977; Jackson, 1974, 1977; Scheltema, 1977) rather than merely the vagaries of sampling and preservation.

An enormous volume of data on the observed ranges of many species exists in the biostratigraphic literature. (I shall refer later to the way in which a pervasive belief in phyletic gradualism has perhaps prevented this information from being adequately appreciated. Here it suffices to point out that the data exist.) We can never be sure that we have really found the earliest or the latest individual of a species. Inferring stratigraphic ranges, however, does not differ in principle from inferring other sorts of ranges. How do we circumscribe the limits of morphological variability? Adult sizes of most organisms conform fairly well to the normal distribution, but even though the normal curve theoretically goes from negative to positive infinity, there is no mouse as big as an elephant. How do we recognize the true geographic boundaries of a species' distribution? Individuals of a species may occasionally be found in places where the species could not have maintained a breeding population; they either may have come there as migrants during a time of temporarily favorable climatic conditions or shifts in current patterns or they may have been displaced after death and transported downslope. In all such cases, presumably, we make informed judgments about whether collecting coverage has been sufficiently complete that the chance of further discoveries making a substantive change in observed range (whether this range be morphological, geographic, or stratigraphic) is "vanishingly small."

As a concrete illustration of this point, I refer to an example of stratigraphic range data collected as part of a study of the systematics of lucinacean bivalves from the Paleocene of the Potomac Valley region (Bretsky and Kauffman, 1977; Kauffman and Bretsky, unpublished manuscript). Figure 4 presents the ranges of seven lucinid species, based on tabulations of species occurrences at 63 stratigraphic horizons within the Brightseat and Aquia Formations. The intervals sampled (defined on the basis of lithological homogeneity with some practical constraints on collecting, such as differences in the degree of induration of beds of different lithology) range in thickness from several centimeters to about 1 m. About 1 $m^3$ of sediment was collected at each horizon, and all identifiable macrofossils were cleaned and counted. We could thus prepare range charts similar to figure 4 for all elements of the macrofauna.

In figure 4, bars connect the earliest and the latest recorded occurrences of a given species in the Potomac Valley Paleocene section; the X's represent species known only from a single horizon. None of the long-ranging species actually occurs at every single collecting horizon within its indicated range; furthermore, if Hazel's (1969) evaluation of the evidence (correlations based on assemblages of ostracodes and foraminifera) of a temporal gap of some 3 to 3.5 million years between Brightseat and Aquia deposition is correct, one has to infer genetic continuity between specimens identified as conspecific

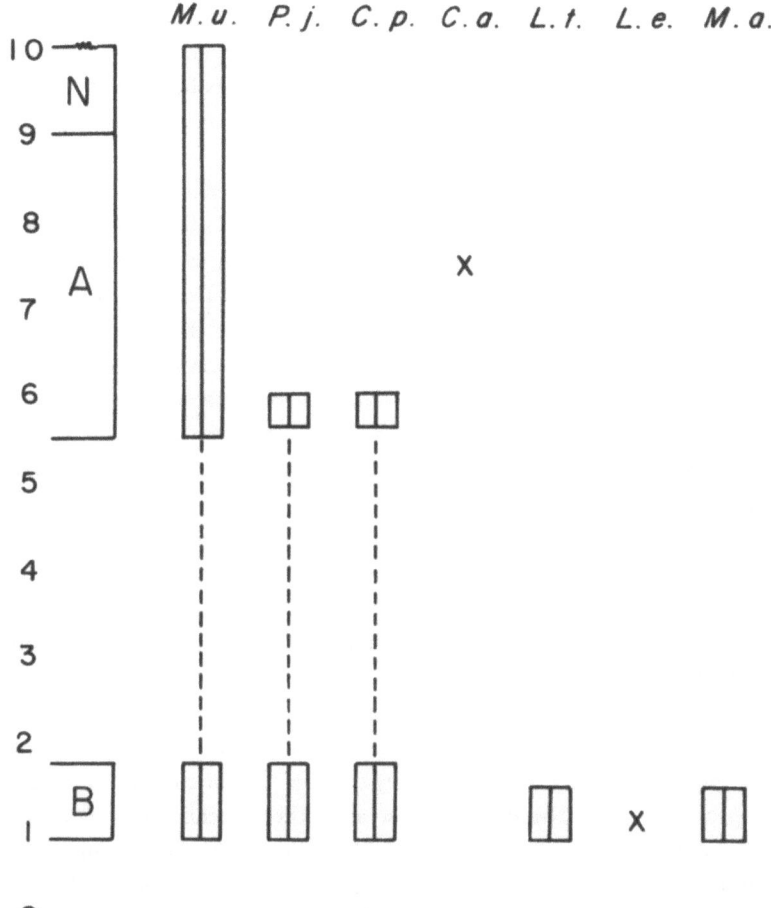

Figure 4. Stratigraphic ranges (see text) of seven lucinid species in the Paleocene and Lower Eocene (part) of the Potomac Valley region. Numbers represent approximate time (in millions of years), with the base of the Paleocene as zero. (Further documentation of chronology will appear in Kauffman and Bretsky, in preparation.) Abbreviations for stratigraphic units: B, Brightseat Formation; A, Aquia Formation; N, Nanjemoy Formation. Abbreviations for species: *M.u.*, *Myrtea uhleri*; *P.j.*, *Phacoides juliannae*; *C.p.*, *Codakia phaedra*; *C.a.*, *Codakia aricia*; *L.t.*, *Lucina temprana*; *L.e.*, *Lucina evansae*; *M.a.*, *Miltha aquiana*.

from the Brightseat and Aquia Formations. On a smaller scale, one also infers genetic continuity between, say, specimens of *Phacoides juliannae* from horizon 1 and horizon 3 of the Brightseat if no specimens of this species have been found at horizon 2. In addition, specimens of the more abundant species belong to a variety of sizes and developmental stages and vary considerably in the quality of their preservation. Thus, even at the level of assigning specimens to species, we are making inferences about genealogical relationships in order to assign specimens from different horizons or localities to a single species which is a genetic continuum.

Two of the Potomac Valley Paleocene species, *Codakia (Codakia?) phaedra* from the Brightseat and Lower Aquia and *Codakia (Cadakia?) aricia* from a single horizon in the Upper Aquia (fig. 4), conform in their stratigraphic relationship to species A and B of Figures 1b and 1c. Details of their sculpture, ligament, and dentition characters indicate that these two species are more closely related to each other than to any other lucinid species known to me. Can we legitimately invoke the observed stratigraphic relationship between these two species to translate their sister-group status into an ancestor–descendant one? Accepting a direct ancestor–descendant relationship between the species (as in fig. 1b, with *C. phaedra* corresponding to species A) implies that their occurrence in the local section does conform to their true order of appearance. Conversely, hypothesizing that both are descended from an unknown common ancestral species (as in fig. 1c) implies that further collecting may be expected to yield *C. aricia* specimens contemporaneous with *C. phaedra* specimens.

In order to decide which of these two alternatives is more likely, we need to look more closely at the details of biostratigraphy and paleoecology of these two species in the Potomac Valley sections. In our collections, there are 28 specimens of *C. phaedra* and most are either juveniles or fragments of adult individuals with partial or entire hinge areas. These specimens come from 11 localities in the Brightseat Formation and 4 localities in the Piscataway Member (Aquia Formation). *Codakia aricia* is represented by only a single, complete, and apparently adult, left valve from the middle Aquia.

The most parsimonious explanation of the morphological similarity between these two temporally successive species would appear to be that *C. phaedra* is directly ancestral to *C. aricia*. Under a strict application of the model of "phyletic gradualism," an ancestor–descendant connection between the two species could only be "proved" by finding an "uninterrupted series of transitional forms at all levels" (Jeletzky, 1965, in Durham, 1967, p. 560). Under the alternative picture of "punctuated equilibria," we are free to hypothesize that the initial population of *C. aricia* was founded as a peripheral isolate (perhaps a successful spatfall in one particular year, when changes in climate or current patterns permitted propagules from the *C. phaedra* parent population to colonize an area which the ancestral species had not previously invaded). The colonization event may have taken place at any time during the "lifespan" of the ancestral species; the apparent stratigraphic "stacking" of the putative ancestor and its presumed descendant might only reflect the distribution of the sedimentary facies characteristically inhabited by each species in the local collecting area.

Looking specifically at the question of habitat preference, the earlier-occurring *C. phaedra* is found most often in an offshore but fairly shallow-water facies, with fewer occurrences in sediments representing more deep water conditions (still no greater than outer shelf depths) included in the Brightseat–Aquia complex of facies. [For further details of sedimentology and paleoecology, see Bretsky and Kauffman (1977) and Kauffman and Bretsky (unpublished ms.).] The single occurrence of *C. aricia* is in the latter, most offshore, facies.

How do these data on environmental occurrence bear on the question of how reliable the observed ranges of the two species in the Brightseat–Aquia section should be considered? The observed duration of *C. phaedra* in the local section (judging from radiometric evidence, to be presented more fully elsewhere) is some 5 to 6 million years, a rather characteristic figure for species longevity in subtidal marine bivalve mollusks (Jackson, 1974; Stanley, 1975). The single occurrence of *C. aricia* is probably a considerable underestimate of the actual longevity of the species, although we cannot exclude the possibility that it was genuinely very short-lived. It may, furthermore, have been so rare that in samples of the size of ours at the various Paleocene horizons (in most cases, a few hundred to a few thousand specimens), the species may fail to be represented simply by virtue of the normal vagaries of sampling error (Dennison and Hay, 1967). We suspect, however, that the failure of the species to show up more commonly in our collections results largely from the fact that in our collecting area, the upper Aquia is predominantly a very nearshore, high-energy sequence. (Our collections cover about one-third of the total area of Brightseat and Aquia exposures in Maryland and Virginia. We chose to concentrate on localities in the Potomac River Valley because these included the type sections of several Paleocene and Early Eocene stratigraphic units and type localities for most of the previously described Brightseat–Aquia species and because collection of many exposures in the suburbs of Washington had to be completed before these were covered over by suburban sprawl.) We predict, therefore, that if geographically more extensive collecting yield better representation of the offshore equivalents of the Potomac Valley Aquia beds, we will obtain additional *C. aricia* specimens.

What is the possibility that older *C. aricia* specimens (whose existence is implied by hypothesizing that both *C. aricia* and *C. phaedra* are descended from an unrepresented common ancestral species) will be found throughout a substantial portion of the range of *C. phaedra?* In the lower part of the Potomac Valley section, *C. phaedra* occurs in facies equivalent to those in which we find *C. aricia* in the middle Aquia. Cooccurrence of two or more lucinid species of the same subgenus is not unknown, but it is less common than the frequent joint occurrence in Recent environments of several more distantly related lucinid species (Allen, 1958; Jackson, 1973; Stanley, 1970; Taylor, 1968). We thus postulate that the failure to find stratigraphically lower *C. aricia* specimens is not an artifact of nonrepresentation of appropriate environments or inadequate collecting effort; rather, the species' absence presumably is a "real" absence. The hypothesis that *C. aricia's* stratigraphic and ecological range comprehends deeper-water shelf facies of the middle and upper Aquia is testable (and potentially falsifiable) by geographically more extensive collecting. Failure to find further representatives of the species, of course, bears all the associated difficulties of proving a negative. Nevertheless, there has to be some (purposely vague) cutoff point to our skepticism about whether failure to find representatives of a particular organism below a particular time plane means that the species actually did not exist before that time. Presumably not even the most doctrinaire phylogenetic systematist would agree with my fundamentalist relatives' interpretation of structures resembling human footprints in the Triassic of Nevada.

In conclusion, we need to look at stratigraphic occurrences in terms of true ranges which can be more or less closely approximated by increased completeness of collecting coverage. Some level of predictability, related to taxonomy, ecology, and biogeography, of average species durations is possible (Jackson, 1974, 1977; Stanley, 1977). Thus we have at least a potential handle on estimating how far off observed durations are likely to be from true durations and thus evaluating how closely the observed chronological relationships between two species conform to their true order of appearance. In the particular case of the two Paleocene *Codakia* species, a good deal of ambiguity remains, but this can be frankly stated, and the assumed ancestor–descendant relationship can be shown to be validated at a modest but not overwhelmingly convincing "confidence level."

## Disjunct Geographic Distribution of Ancestor and Descendant

The preceding discussion touched briefly on the supposedly indispensable link between recognition of ancestor–descendant relationships and acceptance of a model of species transformation by phyletic gradualism. This perceived connection is succinctly expressed in a quotation from Durham (1967, p. 559): "Consideration of the doctrine of evolution and the implications of the biological species concept indicate that our knowledge of the fossil record will not be complete until we have a continuum of morphologies between all taxa with preservable parts." Speaking as phylogenetic systematists, Eldredge (1974) and Cracraft (1974) have likewise assumed an inextricable connection between acceptance of the allopatric speciation model and the tenets of cladistic classification—Eldredge integrating a phylogenetic systematic classification with an ecologically based allopatric speciation model and Cracraft saying that *because* we cannot recognize ancestors, it is impossible to test the model of phyletic gradualism.

As formulated by Eldredge and Gould (1972), the "punctuated equilibrium" model of speciation views new species as originating at the periphery of the ancestral species' distribution [and, by allowable extension, though Eldredge and Gould do not state this explicitly, by chance migration into areas rather distant from the range of the ancestral population, as in the colonization of the Hawaiian Islands by propagules from all points of the compass (Carson, 1970)]. The founders of the new species are expected to be a very small number of individuals, minimally a single gravid female, either highly fecund or with no inhibitions about incest. The speciation event thus is "geologically instantaneous." Transitions between species therefore *do not* generally result from slow, steady, *in situ* morphological transformation of ancestor to descendant; stratigraphic sequence thus cannot be expected to give direct confirmation of proposed ancestor–descendant hypotheses.

Roughgarden (1973, p. 1225), in a comment on Eldredge and Gould's first paper on punctuated equilibria, has expressed some reservations about the conceptualizations of the punctuated-equilibrium model:

> The theory of geographic speciation should be developed to predict *a priori* how strong a barrier must be to produce speciation, how small the peripheral isolate must be, how different the environment must be on each side of the barrier, how long the separation must be maintained, and, especially, how these factors interrelate. Such a theory would predict, where, when, and how fast speciation occurs and would be testable against the phylogenetic record in a much stronger sense than merely providing a new framework for synthesis.

Eldredge and Gould's original definition of the concept of punctuated equilibria (updated somewhat in Eldredge and Gould, 1977) ironically seems to call almost exclusively on a classic, pre–plate tectonics version (Mayr, 1963; Simpson, 1961, 1976) of the allopatric speciation model. For reasons which appeared quite valid before the early 1960s (Hallam, 1973, McKenna, 1973, and Takeuchi, Uyeda, and Kanamori, 1970, provide historical reviews), evolutionary systematists traced routes of diversification and radiation on a temporally stable continental configuration (Mayr, 1976, pp. 549–51; Simpson, 1976). Picturing a restless biota on a stable earth implied that the ancestor of a given species could have lived anywhere at all; given enough time, "improbable chance dispersals become certainties" (Rosen, 1974, p. 289, commenting on Simpson, 1953b).

An alternative to the "peripheral isolate" picture of how allopatric speciation occurs involves, not chance migrations and accumulations of species to make up a fauna [although, as in Hawaii (Carson, 1970), such a situation certainly may obtain in particular cases], but rather the splitting of an entire ancestral biota, and thus of individual species within it, by the emergence of some barrier (see figs. 5 and 6). A familiar example, one much discussed in connection with the possible biological effects of building a sea-level canal (Briggs, 1968; Sullivan, 1977), is the separation and subsequent speciation of marine faunas on either side of the Isthmus of Panama. Keen (1971, pp. 1–2) discusses the effects of the emergence of this particular barrier with reference to its implications for tracing the ancestry of the present-day marine molluscan fauna of the American West Coast.

A similar view of speciation and diversification is familiar to invertebrate paleontologists, especially through Valentine's discussion of diversity change in Phanerozoic shelf biotas, with diversification expected to be greatest at

Figure 5. An example of the "peripheral isolate" variant of the allopatric speciation model. A small founder population (B), derived from a species on the western island (A), migrates (arrow) to the eastern island. It is ancestral to both of the closely related species (C and D) on the eastern island. (After Carson, 1970; copyright 1970 by the American Association for the Advancement of Science.)

times of greatest continental fragmentation (Valentine, 1969, 1977; Valentine and Moores, 1970). Hennig's concept of vicariance (Hennig, 1966; Croizat, Nelson, and Rosen, 1974; Rosen, 1974) is comfortably assimilated into plate-tectonic reconstructions. Cracraft (1973, 1974) and Platnick (1976), among others, have shown that many distributional patterns in terrestrial faunas are similarly best understood as the results of vicariance in response to changing continental geography rather than as the cumulative effects of chance dispersals.

Although phylogenetic systematists have not, to the best of my knowledge, explicitly attempted to answer questions about the speciation process like those asked in Roughgarden's review of Eldredge and Gould's paper (Roughgarden, 1973), the vicariance picture of allopatric speciation clearly contrasts with the peripheral isolate one in implying that the two descendant populations of an ancestral species will each be relatively large and wide-

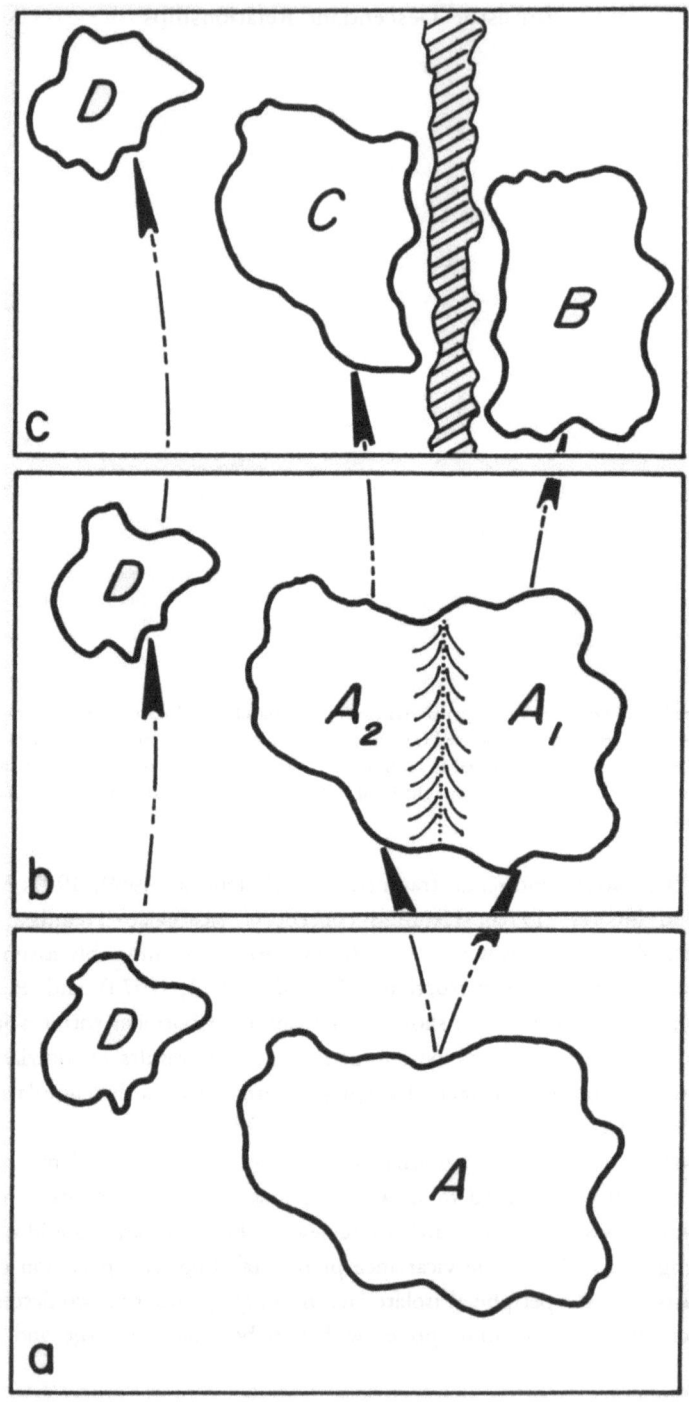

spread. Each population potentially incorporates a substantial proportion of the ancestral species' total variability. The speciation event will not necessarily be geologically instantaneous, and independent paleogeographic evidence may permit us to place an approximate date and location on the origin of particular species (Bretsky, 1975; Eldredge and Gould, 1977). A strict distinction between "phyletic gradualism" and "punctuated equilibria" thus breaks down. The impetus to speciation is provided by the "event" (often perhaps a fairly lengthy one) which divides the parent population, but *in situ* accumulation of species-level differentia may also occur after the subdivision of the parent population (Bretsky, 1975; Sylvester-Bradley, 1977). Insistence that speciation events must necessarily involve small populations (Gould, 1977, p. 322, and personal communication, 1977) appears to be overly dogmatic. Allowing for a certain component of taxonomic frailty, the vicariant East Pacific and Caribbean–West Atlantic marine mollusk species (Keen, 1971) or numerous sister species of terrestrial arthropods and vertebrates (like those discussed by Platnick, 1976, and Cracraft, 1973) *are* good species, and presumably, given enough time for further speciation events to occur, may become (or may have been) the founders of higher-level taxonomic groups. But in the vicariance model, species-level differentiation is not achieved by the separation of a very small peripheral isolate and subsequently reinforced by the perfection of mechanisms to guarantee reproductive isolation (Eldredge and Gould, 1972; Mayr, 1963, 1976, pp. 129–34), for the descendant species may never come back into contact with the ancestor.

In summary, we encounter a paradoxical situation. Evolutionary systematists who have wanted to recognize ancestor–descendant relationships have also accepted a portrayal of speciation events as occurring strictly in very small peripheral isolates. This "floating crap game" model of dispersal and speciation implies that distributional evidence for ancestor–descendant relationships will be hard to come by. The ancestor of a particular species may not have come from within the same or a contiguous basin of deposition, but

Figure 6. An example of the "vicariance" variant of the allopatric speciation model. Arrows indicate ancestor–descendant relationships between populations from different times. (*a*) Original distribution of two species, one wide-ranging (A), and one more restricted (D). (*b*) An early stage in the construction of a barrier which separates A's distribution into two partly isolated populations, $A_1$ and $A_2$. (*c*) Situation after completion of the barrier and species-level differentiation of two descendants, B and C, of the earlier species A. (After Bretsky, 1975.)

by chance migrations from anywhere at all. Phylogenetic systematists, by contrast, have advocated recognizing only sister-group relationships. But they have adopted a picture of speciation events as being patterned in their occurrence [e.g., Platnick's (1976) demonstrations of a close correspondence between the sequence of continental fragmentations and the hierarchy of sister-group relationships among laroniine spiders], involving relatively large-sized descendant populations and requiring a fair amount of time for completing the barrier to dispersal between vicariant species. Under the vicariance model, it is not necessarily true that "small numbers and rapid evolution [will] virtually preclude the preservation of speciation events in the fossil record" (Gould and Eldredge, 1977, p. 117). One may thus argue that the vicariance model actually implies a more optimistic view of the possibility of recognizing ancestor–descendant relationships between particular species than the peripheral isolate model of speciation.

## Morphological Discontinuity between Ancestor and Descendant

I have already remarked that in a vicariance speciation event, each of the two descendant populations may be expected to incorporate a substantial fraction of the total genetic and phenotypic variability of the ancestral species. This expectation points up an apparent contradiction between the vicariance picture of speciation events and the phylogenetic systematic model (as expressed, for example in Hennig, 1966, fig. 21) of speciation and character transformation. To reiterate briefly, this model (fig. 7) states that in any speciation event, at least one character is changed from the ancestral to the derived state. [Obviously, for a speciation event to be recognized as such, any speciation model requires diagnostic change in at least one character, but evolutionary systematic portrayals of speciation generally allow for the possibility of ambiguity in the species assignment of single individuals (e.g., Gould, 1969, p. 467, pl. 2. fig. 5).] The phylogenetic systematic model further states that subsequent speciation events can involve either additional change in that same character, which results in a morphocline from ancestral to most highly derived character state (fig. 7b), or change in different characters (fig. 7c).

# Ancestor–Descendant Relationships 143

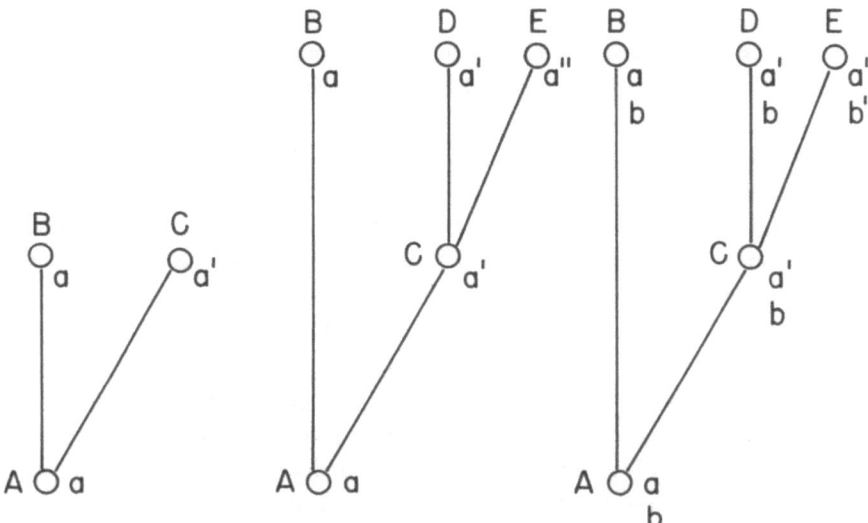

Figure 7. Phylogenetic systematic model of speciation and character divergence (see text). A, B, C, D, and E are species; a, a′, and a″ are, respectively, ancestral and progressively more derived states of one character; b and b′ are ancestral and derived states of a second character. (After Hennig, 1966.)

The phylogenetic systematic model tacitly presents species as monomorphic for either the ancestral or the derived state of a diagnostic character; it does not explicitly deal with the notion of morphological variability within the ancestral species. [This is not to say that individual phylogenetic systematists are indifferent to the existence of variability. Eldredge's work (1972, 1973, 1974), for example, provides some extremely elegant examples incorporating intraspecific variation into a cladistic analysis. The point is just that few diagrammatic portrayals (as opposed to what is done in actual practice) in cladistic literature of speciation events attempt to deal with variation, and as Eldredge and Gould (1972) have emphasized in another context, pictures do color perceptions.]

The enthusiastic response of some phylogenetic systematists (e.g., Cracraft, 1974) to the "peripheral isolate" variant of the allopatric speciation model may result partly from the fact that this formulation of the concept of "punctuated equilibria" predicts (and explains) decided morphological gaps between ancestral and descendant species (which, because no morphological transitions exist and stratigraphic evidence is unreliable, can be recognized only as sister species on phylogenetic systematic tenets). If most speciation occurs via migration of a propagule from the geographic periphery of the ancestral species, the individuals involved will probably be genetically and phenetically distinctive already (Eldredge and Gould, 1977, p. 35). Migration of such a propagule into an area not occupied by conspecifics or close competitors may also permit a "population explosion" which, when the eventual "crash" occurs (Carson, 1975), ends with a descendant species composed of individuals differing quite substantially from the ancestral population. Such a population "flush" and "crash" may occur on a geologically, and even ecologically, instantaneous time scale.

Carson's model appears to be quite successful in explaining, for example, ancestor–descendant relationships (through a correlated evaluation of chromosomal, morphological, biogeographic, and geological evidence) within taxa such as the Hawaiian Drosophilidae. From a practical standpoint, a speciation event which begins with a very small founder population and in which species-level differentiation in at least one diagnostic character is attained very rapidly may be operationally indistinguishable from the sort of "phase change" transformation of species A, monomorphic for plesiomorphous character state $a$, into species C, monomorphic for apomorphous character state $a'$, diagrammed in figure 7. Nevertheless, the concept of species as intrinsically variable [stated most vividly, perhaps, in the context of human racial classification by Simpson (1969, pp. 98–106)] is one of the most important contributions of Simpsonian–Mayrian evolutionary theory and should not be lightly discarded, particularly now that most systematists have become comfortable with quantitative techniques permitting a more objective assessment and communication of patterns of variability. Whatever their theoretical shortcomings or practical inadequacies, "textbook" examples of phyletic gradualism (e.g., Trueman, 1922, fig. 5; reproduced as fig. 5-1 of Eldredge and Gould, 1972) do, at least, generally present successive fossil populations as inherently variable, with one segment of the total range of ancestral variability being favored by selection to produce the descendant.

An overly enthusiastic synthesis of punctuated equilibria and cladistic classification risks throwing the variational baby out with the gradualistic bathwater. The alternative to insensibly graded, stratigraphically stacked morphological series is not necessarily a complete dichotomy between ancestor and descendant for a given character. "Stasis within species" still potentially subsumes a great deal of intraspecific variability, though chiefly in the lateral rather than in the vertical dimension. If ancestral species are perceived as inherently variable, then a strict dichotomy between ancestral and derived character states is meaningless, for characters which are fixed in the descendant species may have existed as variants within the ancestral population. Thus, if a species stratigraphically positioned so as to be a possible ancestor of a particular taxon diverges slightly in one or more characters from the hypothetical ancestral species (defined in terms of a summation of plesiomorphous characters) of that taxon, then, contrary to statements by Farris (1976, p. 272) and Tattersall and Eldredge (1977, p. 207), the expectation of intraspecific variability means that such a discrepancy need not preclude interpretation of that species as an actual ancestral species. Since all specimens of the ancestral species are not identical, no single specimen or small collection of specimens can be expected to correspond exactly to a hypothetical ancestral morphotype. One is uncomfortable at making such a statement, for it seems to reintroduce into phylogenetic reconstruction the subjective element of "art" which both phylogenetic systematists and pheneticists have criticized in the evolutionary systematic model. Yet results never correspond exactly to predictions in any science, and in phylogeny reconstruction, as well as in physics or theoretical ecology, one component of scientific creativity consists in knowing what discrepancies to ignore.

In this connection, the vicariance picture of allopatric speciation puts an even stronger emphasis on the assessment of intraspecific variability than does the peripheral isolate portrayal. The vicariance view implies rather substantial size of both fractions of a subdivided ancestral species and, thus, initially no absolute distinction between ancestral and descendant morphologies. Figure 8 provides a simplistic, but perhaps heuristically useful, diagram to illustrate this point. If the distribution of a species which shows clinal variation in some character is subdivided by a barrier trending perpendicular to the direction of the cline, the descendant populations will differ, on the average, from each other and from the ancestral species. But in neither descendant species (at least early in the history of their differentiation) is there a "derived" state of this character which in an absolute sense differs from the "ancestral" condition.

Further evidence against the assumption of a strict dichotomy between "ancestral" and "derived" conditions comes from an old, but newly respectable, view of how major morphological changes may come about rather rapidly. Classical evolutionary systematists presumably would not disagree with that aspect of the phylogenetic systematic model of speciation and character

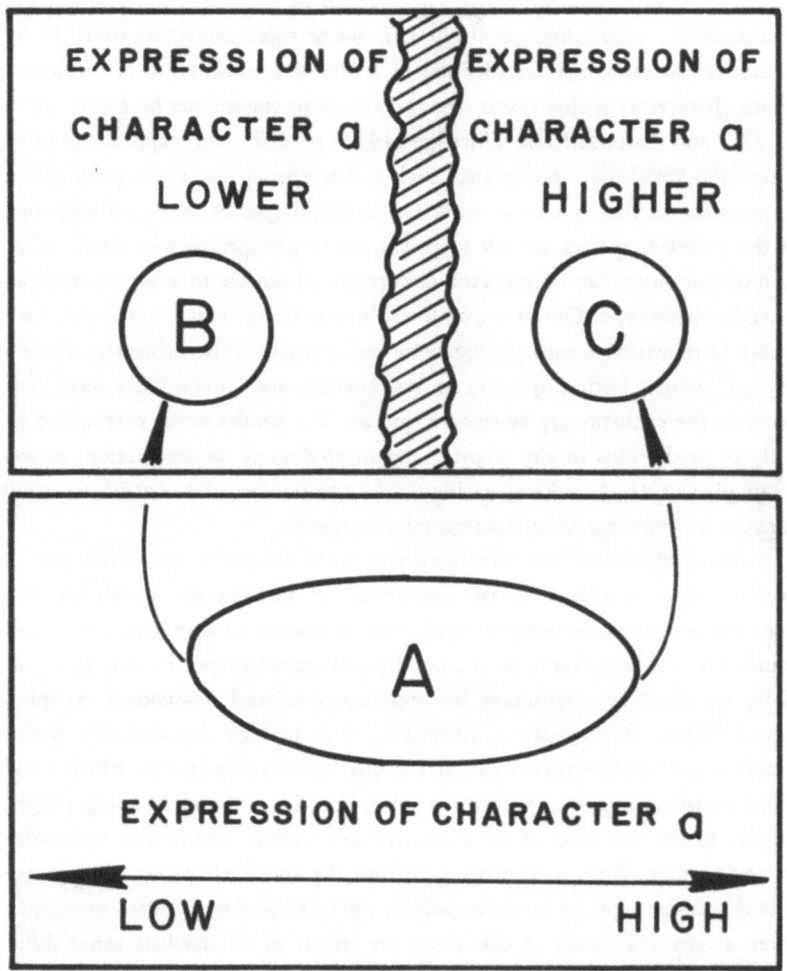

Figure 8. The vicariance model and clinal variation. A (see fig. 6) represents a species showing clinal variation in the expression of some character (a), from low in the west to high in the east. If the distribution of species A is divided by a barrier and speciation is carried to completion, descendant species B will have, on the average, a lower value, and descendant species C will have, on the average, a higher value for that character. Such a variable character may be a valuable indicator of phylogenetic affinity even though no state of the character can be said to be "ancestral" or "derived."

divergence (fig. 7) which presents macroevolutionary events as an accretion of step-by-step changes in single characters. Indeed, evolutionary systematists (see particularly Simpson, 1953a, pp. 338–76) have strongly stressed the continuity between microevolution and macroevolution. But recently, Frazzetta (1970, 1975) and Valentine and Campbell (1975), among others, have perceived major morphological innovations as resulting from rather small changes in the timing of developmental events. The variability that is present and potentially available to result in new morphological designs exists within ontogeny, as well as among cooccurring adults (as in the vicariance–clinal variation diagram of fig. 8). Frazetta (1970) jokingly summarizes his view of some of the major features of evolution as rehabilitating the "hopeful monster." Gould (1977) enlarges on this theme to demonstrate how adaptive innovations may frequently have originated as changes in reproductive and developmental strategies—changes occurring initially only to obtain short-term advantage in fluctuating or in stable environments but, in a small minority of cases, having had much greater evolutionary significance.

When one turns to working out the phylogeny of a specified group of organisms, the view that small changes in developmental timing may potentially result in rather large differences in morphology implies that we must look at organisms as morphologically integrated wholes. By contrast, the traditional cladistic strategy of phylogeny reconstruction seems to treat organisms as simply the additive result of a combination of independent character states. Choice among competing cladograms has been made essentially on the basis of majority vote—Which possible cladogram involves the fewest reversals and parallelisms? (This question involves an assessment, admittedly partly speculative, of which states of the characters concerned are ancestral and which derived; see Eldredge, in this volume). Little consideration appears to have been given to the possibility that a rather small evolutionary "event" could have resulted in relatively great morphological divergence involving characters which are not independent, but highly correlated. As with the question of recognizing intraspecific variability, practice has here outstripped formalized principle. Embracing phylogenetic systematics has not inhibited Eldredge (1968, 1972) and Cracraft (1976), for example, in carrying out quantitatively sophisticated studies of morphological integration. But no one yet has proposed a model of phylogeny reconstruction based on cladistic principles which explicitly incorporates these contributions of developmental biology and multivariate morphometrics.

A brief look at the details of morphology of the two Paleocene species of *Codakia* whose stratigraphic relationship was discussed earlier (see fig. 4) exemplifies the point that even a less rigorous consideration of character correlation and ontogenetic change may have value as an aspect of phylogeny reconstruction. Adult specimens of the stratigraphically lower *Codakia phaedra* attain about twice the size of the, apparently adult, single specimen of *C. aricia*. As adults, the two species differ in several morphological features. In both, the first 2 mm of the shell below the beak have only concentric elements of sculpture. In *C. phaedra*, however, the later-developing radial elements of sculpture are less prominent than the concentric ones, producing a "frilled" surface ornament, whereas in *C. aricia*, radial and concentric elements of sculpture are about equally prominent, producing a cancellate ornament. The ratio of anterior length (length of the shell anterior to the beak) to total length is greater in *C. aricia* than in adult *C. phaedra*, and the ligament of *C. aricia* is set in a shallower groove and is closer to the posterodorsal margin of the shell than is that of adult *C. phaedra*. In these latter two characters, juveniles of *C. phaedra* are more similar to *C. aricia* than to adults of their own species. It is thus plausible to interpret the differences in shell shape and placement of the ligament in adults of the two species as consequences of accelerated maturation of *C. aricia* relative to *C. phaedra*. Similar ontogenetically correlated differences occur in many other lucinid species as well. [See, for example, Bretsky (1976, p. 282), wherein I have interpreted a supposedly diagnostic difference in lunule shape of two other species of *Codakia*—the Recent Atlantic–Pacific species pair *Codakia orbicularis* and *Codakia distinguenda*—as a simple consequence of the greater size of *C. distinguenda* by extrapolating to greater size ontogenetic changes occurring in the smaller species.]

Obviously, recognizing morphological differences between species as correlated with size differences neither gives an unambiguous delineation of the polarity of ancestor–descendant relationship between two sister species nor eliminates the possibility of both species being descended from an unknown common ancestor. But the recognition that fairly substantial morphological differences among adults of closely related species may result from rather minor changes in timing of ontogenetic events and that these shifts may go in either direction (acceleration or retardation of ontogenetic development), depending on what provides short-term ecological advantage (Gould, 1977, pp. 285–94), militates against the possibility of unambiguously recognizing "ancestral" and "descendant" character states and the expectation of unidirectional morphoclines. Evolutionary reversals may be extremely common. Thus stratigraphic position, assessment of overall similarity, and analysis of the ecological significance of diagnostic character differences between similar species (in terms of both functional morphology and life-history strategy) may be more productive guides to phylogenetic relationship than is *a priori* assessment of morphocline polarity, allowing for little or no evolutionary reversal.

As a final point in this discussion, intraspecific morphological variability can be expressed in dichotomous or multichotomous states of qualitative characters, as well as in continuous variation of quantitative ones. Characters associated with sexual dimorphism are an obvious example of this sort of qualitative intraspecific variability. How recognizing the existence of sexual dimorphism in fossil specimens can reorient systematics has been demonstrated for organisms as different as dinosaurs (e.g., Dodson, 1976) and ammonites (e.g., Kennedy and Cobban, 1977). Kauffman and Buddenhagen (1969) discuss an illustrative example of apparent sexual dimorphism in fossil mollusks, specifically astartid bivalves. In the Paleocene samples which they studied, all specimens smaller than 18 mm in length lack crenulations on the ventral commissure, whereas about 70 percent of those larger than about 20 mm in length have them. In living *Astarte*, which are protandric (changing from male to female at about one-half to two-thirds full adult size), "large adults are predominantly females and predominantly crenulate in about the same proportions" (Kauffman and Buddenhagen, 1969, p. 82). It thus appears legitimate to infer that a similar relationship holds for the fossil specimens and that what could potentially be regarded as a diagnostic character discriminating a large, crenulate species from a small, noncrenulate one actually represents a biologically significant source of intraspecific variability.

Hennig (1966, p. 74) and other phylogenetic systematists have strongly criticized the use of "overall resemblance" as a guide to the relationship of species. Yet at the level of sorting and identifying species in a sample of fossil invertebrates, one begins with a visual assessment of overall similarity, with no distinction between "ancestral" or "derived" character states, in order to make the initial groupings. One may then bring in evidence and arguments, analogous to those which Kauffman and Buddenhagen used to recognize the two Paleocene *Astarte* morphs as conspecific, in order to infer genetic continuity among highly similar, yet somehow different, specimens. If this procedure is legitimate at the species level, then given documentation of adequate stratigraphic coverage and adequate representation of species diversity (as discussed in preceding sections of this paper), the procedure can be extended to inference of ancestor–descendant relationships between similar species with closely successive or overlapping stratigraphic distributions. The vastly over-simplified model of speciation and character transformation (fig. 7) which portrays each species as absolutely monomorphic for either the ancestral or the derived state of a given character must be revised to correspond more closely to ecological and evolutionary reality if phylogenetic systematics is to gain widespread acceptance.

## Ancestor–Descendant Relationships between Supraspecific Taxa

The preceding discussion has explored a model of speciation and character change which rejects the necessity of phyletic gradualism but retains the importance of considering morphological variability within species. By breaking down a perceived dichotomy between recognizing genetic continuity among conspecific organisms (organisms viewed as monomorphic for ancestral or derived states of diagnostic characters) and recognizing such continuity between different species (organisms viewed as absolutely discrete entities with respect to which state of such characters they possess), I have tried to indicate that a speciation model which synthesizes the insights of vicariance, intraspecific variability, and the importance of ontogenetic change removes some of the apparent objections to inferring ancestor–descendant relationships between species.

Species vary, and so do higher taxa. Cracraft (1974) and Bonde (1975) have been among the most articulate of phylogenetic systematists in arguing that ancestors at any rank, not just ancestral species, "cannot be recognized or identified but can only be hypothesized" (Cracraft, 1974, p. 73). These workers reject the concept of "minimal monophyly" (Simpson, 1961), formulated to permit expression of the inference that, for instance, a particular genus is ancestral to another when stratigraphic evidence is inadequate for establishing the exact network of connections between the species included in the two taxa.

Because any nonmonotypic supraspecific taxon has a diagnosis which is broader and more vague than the diagnosis of any of its included species and because there is no "satisfactory general measure of evolutionary differentiation" (Griffiths, 1973, p. 338), ranking in the taxonomic hierarchy has traditionally been used as a tacit summarization of the level of variability within a higher taxon as well as the degree of morphological divergence from other taxa. [See Valentine (1969) and Sprinkle (1976) for comments on the ecological justification for high-level taxonomic separation of taxa of low diversity, early Phanerozoic origin, and noteworthy morphological distinctiveness.]

Workers who have used the products of systematic research to investigate whether or not there are discernible patterns in the course of evolution appear to have implicitly regarded degree of morphological differentiation, rather than strict monophyly, as the criterion for defining higher taxa. Some examples are Flessa and Imbrie's (1973) essay documenting periods of coincident evolutionary activity in various marine and terrestrial groups, Boucot's (1975) comparison of rates of production of new genera in cosmopolitan and provincial brachiopods, and Gould et al.'s (1977) analysis of characteristics of randomly generated and "real" clades ("real" as defined by sources such as the *Treatise on Invertebrate Paleontology*). Farris (1977) has legitimately criticized an early presentation by Gould (1975a) of the results of this last study for using terms with phylogenetic systematic implications in describing clearly paraphyletic groups; however, the intent of Gould et al. (1977) was to use criteria for defining taxa among their simulated organisms which corresponded to criteria used by the majority of taxonomists.

In evolutionary systematic practice, finding species which appear to overlap the morphological boundary between previously defined genera (or other taxa) has been taken as evidence of close relationship between the higher taxa involved. When paleontological data have been available, stratigraphic sequence has further been used as an indicator of the direction of ancestor–descendant relationship between the taxa (though often with reservations when the apparently transitional species occurs in what appears to be the "wrong" temporal position).

As a brief illustration of how variation within and among higher taxa is used to infer ancestor–descendant relationships between such taxa, I cite an instance from my work on the phylogeny of Cenozoic Lucinidae. Within the genus *Lucina (sensu lato)*, three subgenera with Miocene to Recent records—*Parvilucina, Radiolucina* (formerly *Bellucina,* in part), and *Pleurolucina*—represent a progression from more to fewer radial ribs. I had previously (Bretsky, 1969, 1976, fig. 3) interpreted this progression as a directed morphocline.

*Pleurolucina,* but not the other two taxa, shares with another subgenus, *Lucina (sensu stricto)*, a rare state (whether ancestral or derived, I do not know) of the character "lunule shape." With this resemblance as one impetus, Britton (1972) explored an alternative hypothesis of closer relationship between *Pleurolucina* and *Lucina s.s.* than between *Pleurolucina* and the other two radially ribbed subgenera. A major portion of Britton's study involved analyzing the ontogeny of several Recent species of both *Pleurolucina* and *Lucina s.s.* He showed that the sulci which define radial ribs in *Pleurolucina* also appear in the early ontogeny of species whose dentition, musculature, and elements of concentric sculpture correspond to those of the type species of *Lucina s.s.* and other species unambiguously assigned to that subgenus. He also presented evidence that the details of development of radial ribbing in *Pleurolucina* differ sufficiently from the condition in *Parvilucina* and *Radiolucina* that the character state "radial ribbing" should actually be divided to reflect the occurrence of two rather different modes of attainment of this sort of shell sculpture.

I concur with Britton in his evaluation of the evidence for close relationship between *Pleurolucina* and *Lucina s.s.* This affinity may be expressed either in terms of ancestor–descendant relationship between the two subgenera or in terms of an unknown common ancestor for the two. The parent–daughter relationship is a much more narrowly defined concept than that of the sister group; the latter relationship deals only with "those two . . . taxa *among all those known to the systematist* which most recently shared a common ancestry" (Eldredge and Tattersall, 1975, p. 225; italics theirs). But for a taxon such as Cenozoic bivalves, it is less realistic to postulate totally unrepresented subgenera (or higher taxa) than it is to postulate totally unrepresented species.

Furthermore, when we consider the stratigraphic relationship between *Pleurolucina* and *Lucina s.s.,* the latter (figure 9) is recorded from middle Eocene time onward (Bretsky, 1976, Conrad, 1833; Dall, 1903), whereas the first record of *Pleurolucina* is of probable early Miocene age (an occurrence

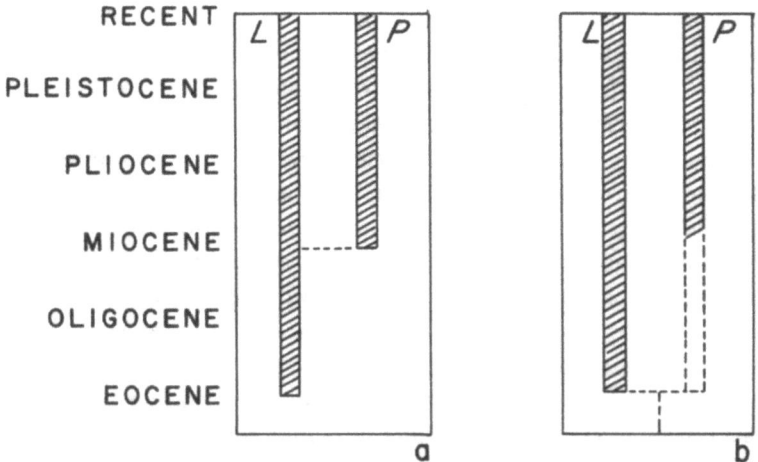

Figure 9. Known stratigraphic records of *Lucina* (L) and *Pleurolucina* (P), with two different interpretations of phylogeny, and their implications about actual temporal ranges. (*a*) Derivation of *Pleurolucina* from an unknown species of *Lucina*; known stratigraphic records of both subgenera taken as corresponding approximately to true total ranges. (*b*) Derivation of *Lucina* and *Pleurolucina* from unknown species of an unknown subgenus; unrecorded portion (dashed lines) of total stratigraphic range of *Pleurolucina* taken as rather greater than unrecorded portion of total range of *Lucina*.

in Ecuador, reported by Olsson, 1964). This early Miocene species, like some of the Recent species of *Lucina s.s.* discussed by Britton (1972), has well-marked radial ribs only in early ontogeny, but its concentric sculpture pattern is more like that of typical *Pleurolucina* than that of typical *Lucina s.s.* Given this stratigraphic distribution, stating simply that *Pleurolucina* and *Lucina* have an unknown common ancestor implies that both taxa may have originated as far back as early Eocene time but that fossil representatives of *Pleurolucina* were not preserved for a substantial portion of the duration of the subgenus. If data about the temporal duration of supraspecific taxa are valid (with the routine paleontological caveats) for studies of such questions as patterns of diversification and faunal resemblance among biogeographic provinces, it is not particularly consistent to consider them useless for evaluating the direction of ancestor–descendant connections between taxa.

Harper (1976, p. 184) has previously pointed out that the likelihood of particular fossil taxa being directly ancestral to others in a given lineage "might well be low for the species level, but would increase rapidly with the taxonomic level considered." I would extend his argument to contend that expressing an ancestor–descendant relationship between supraspecific taxa of a particular rank symbolizes the degree of confidence which one has in a hypothesis about phylogeny, much as one uses statistical confidence intervals to convey the reliability of an estimate of a population parameter. In the case of *Lucina* s.s. and *Pleurolucina*, I find it most realistic to express my level of confidence in Britton's hypothesis of their relationship by stating that *Lucina* s.s. is ancestral to *Pleurolucina* (a statement consistent with both ontogenetic evidence and stratigraphic position, taking into account the variability which exists both among and within species of the two subgenera and the variability in ontogeny among individuals of these species) but that we are unable to identify the exact transitional species.

Although, formally, an expression of ancestor–descendant relationship at the subgeneric level may seem to differ little from simply postulating an unknown common ancestor for *Pleurolucina* and *Lucina* s.s., several systematists have argued that specific statements about the direction of ancestor–descendant relationships are to be preferred because they are firmly rooted in a particular time, place, ecological setting, and complex of selection pressures (Bretsky, 1975; Gould, 1975b; Ross, 1974). As Gould (1975b, p. 245) phrases the argument, "Better a close relative of an ancestor in all the fullness of its anatomy than a short list of necessary characters for the unknown ancestor itself."

Reconstruction of phylogeny is a valid aim in itself, and one does not want to justify the existence of invertebrate paleontology as the handmaiden of ecology rather than of stratigraphy. Nevertheless, many interesting evolutionary questions can be approached only if one is willing to specify a particular direction of ancestor–descendant relationship. Such a question has been raised at the species level as follows (Jared Diamond, written communication, 1977): "An important question in interpreting the apparent turnover [appearances and disappearances of species, as in table 2] is whether it consists of population lines actually disappearing from an area, new lines invading from other areas, and old lines persisting in an area but speciating to yield two sympatric lines (i.e., real extinction, real immigration, and real speciation, respectively); or whether it simply consists of morphological change [via phyletic gradualism] without speciation (speciation defined as splitting of one line to yield two)."

Evolutionary systematics has often been criticized for the vagueness with which both its methodology and its results seem to be expressed. Both pheneticists and phylogenetic systematists have reacted strongly against Simpson's (1961, p. 107) candid statement about the component of "art" in systematics. In a recent thoughtful attempt to formalize more precisely the process of phylogeny reconstruction, Eldredge and Tattersall (1975; Tattersall and Eldredge, 1977) have suggested a process of analysis in which one proceeds from a cladogram (a simple branching pattern based on the distribution of shared derived characters) to a tree (a diagram, often obtained in part by superimposing a cladogram on a stratigraphic section, in which ancestor–descendant relationships are specified) to a scenario (a narrative which attempts to explain a tree in terms of selection pressures, interspecific competition, changing environments and biogeography, and so on).

Unfortunately, Tattersall and Eldredge (1977) have partly undercut their own persuasive argument by their choice of an example of a "scenario." Most professional paleontologists would justifiably resent an imputation that their work is based on as shallow a comprehension of ecology and genetics, not to mention stratigraphy and sedimentology, as the work they cite (Morgan, 1972) demonstrates. Much of the excitement of invertebrate paleontology, particularly in the past 10 years or so, has indeed proceeded from attempts to look at adaptation and evolution in terms of just those concepts and kinds of information which are involved in serious efforts at Eldredge and Tattersall's "scenario" level of phylogeny reconstruction. In practice, we often find that we cannot recognize particular characters as homologous or nonhomologous unless we know something about the organism's relationship to its environment. Inferences about morphocline polarity based simply on comparative morphology separated from a consideration of environmental context may prove futile or at least no more productive than similar inferences based on information from the stratigraphic sequence of taxa.

Phylogenetic systematics, especially in its extreme formulation, can be expected to encounter resistance from practitioners of invertebrate paleontology on at least two levels. Professionally, it may be perceived as a reactionary movement, negating the training of students to look at fossil organisms not as static morphologies out of museum drawers or sent in by a field geologist for identification but as once-living creatures molded by the demands of a particular environment. In a more personal context, it may have been premature to celebrate the demise of reconstructions of ancestor–descendant relationships (even if these admittedly contain a component of symbolic rather than literal truth) in the same year that 130 million people watched "Roots."

## Summary

Fossil species cannot be adequately symbolized by equal-sized black dots occurring at single stratigraphic levels and monomorphic for particular character states; they have stratigraphic ranges, geographic distributions, and vary morphologically on several scales—within individual ontogeny, among localities, and perhaps through time as well. All taxonomy, including species-level taxonomy, thus involves more or less well-founded inferences about genetic connections between populations (both statistical and biological). Inferring ancestor–descendant connections between species or between supraspecific taxa thus calls on the same sorts of evidence and lines of arguments as those which are used to recognize morphologically distinct individuals as conspecific. We cannot compartmentalize our efforts by saying that studying morphological variability belongs only to the stage of species identification or that information on ecological setting is not brought in until the scenario level of phylogenetic analysis. Not all stratigraphic data are equally reliable. The nature of the organisms being studied, the precision with which temporal correlations can be made, and the level at which one wishes to infer ancestor–descendant relationships all affect the reliability of stratigraphic data as a guide to such relationships.

## Acknowledgments

For helpful comments and discussion at various stages of preparation of this manuscript, I thank Peter Bretsky, Karl Flessa, Roger Thomas, Scott Brande, A. R. Palmer, Stephen Gould, Jared Diamond, Joan Miyazaki, Philip Signor, Charles Harper, Thomas Schopf, J. S. Farris, and E. O.

Wiley. These people do not necessarily agree with me, and some may not even remember what they have said to inspire or to provoke me. I am grateful to them nevertheless. I thank Joel Cracraft and Niles Eldredge not only for having invited me to participate in their symposium but also for having assured me of freedom to say or to write anything I believed. Our common ground has been that of clarifying issues rather than converting unbelievers. I also thank Erle Kauffman for having facilitated my collaboration on a project which forced me to think in more concrete ways about issues which previously had been only theoretical constructs. Our work on the Paleocene Potomac Valley lucinaceans was partly financed by the National Science Foundation (Grant GB-27441), the Smithsonian Research Foundation, and the State University of New York at Stony Brook. Finally, I thank Edward Marintsch for his suggestions for locating obscure references and Joyce Zummo and Philip Bretsky for cooperating with me and with each other under often difficult conditions, including the worst winter in the northeastern United States in 177 years, while I worked on this manuscript.

# References

Allen, J. A. 1958. On the basic form and adaptations to habitat in the Lucinacea (Eulamellibranchia). *Phil. Trans. Roy. Soc. London Ser. B*, 241:421–84.

Bambach, R. K. 1977. Species richness in marine benthic habitats through the Phanerozoic. *Paleobiology* 3:152–67.

Blackwelder, B. W. 1972. Morphometry, evolution, and phylozones of the molluscan genus *Cavilinga* (Bivalvia; Lucinidae) in the Pliocene to Holocene of the southern Atlantic Coastal Plain. Ph.D. thesis, George Washington University.

Bock, W. J. 1973. Philosophical foundations of classical evolutionary classification. *Syst. Zool.* 22:375–92.

Bonde, N. 1975. Origin of "higher groups": Viewpoints of phylogenetic systematics. *Colloq. Int. CNRS* 218:293–324.

Boucot, A. J. 1975. *Evolution and Extinction Rate Controls*. Elsevier, Amsterdam.

Bretsky, S. S. 1969. Phenetic and phylogenetic classifications of the Lucinidae (Mollusca, Bivalvia). Ph.D. thesis, Yale University.

Bretsky, S. S. 1971. Evaluation of the efficacy of numerical taxonomic methods: An example from the bivalve mollusks. *Syst. Zool.* 20:204–22.

Bretsky, S. S. 1975. Allopatry and ancestors: A response to Cracraft. *Syst. Zool.* 24:113–19.

Bretsky, S. S. 1976. Evolution and classification of the Lucinidae (Mollusca; Bivalvia). *Palaeont. Amer.* 50:217–337.

Bretsky, S. S., and Kauffman, E. G. 1977. Morphological variability and temporal change in a Paleocene lucinid bivalve mollusk. *Bull. Geol. Soc. Denmark* 26:161–74.

Briggs, J. C. 1968. Panama's sea-level canal. *Science* 162:511.

Britton, J. C., Jr. 1970. The Lucinidae (Mollusca: Bivalvia) of the Western Atlantic Ocean. Ph.D. thesis, George Washington University.

Britton, J. C. 1972. Two new species and a new subgenus of Lucinidae (Mollusca: Bivalvia), with notes on certain aspects of lucinid phylogeny. *Smithsonian Contr. Zool.* 129:1–19.

Carson, H. L. 1970. Chromosome tracers of the origin of species. *Science* 168:1414–18.

Carson, H. L. 1975. The genetics of speciation at the diploid level. *Amer. Nat.* 109:83–92.

Cifelli, R. 1969. Radiation of Cenozoic planktonic Foraminifera. *Syst. Zool.* 18:154–68.

Colbert, E. H. 1961. *Evolution of the Vertebrates.* Science Editions, New York.

Conrad, T. A. 1833. Fossil Shells of the Tertiary Formations of North America, No. 2. Privately printed, Philadelphia.

Cook, H. E., and Taylor, M. E. 1975. Early Paleozoic continental margin sedimentation, trilobite biofacies, and the thermocline, western United States. *Geology* 3:559–62.

Cracraft, J. 1973. Vertebrate evolution and biogeography in the Old World tropics: Implications of continental drift and palaeoclimatology. In D. H. Tarling and S. K. Runcorn (eds.), *Implications of Continental Drift to the Earth Sciences,* Vol. 1, pp. 373–93. Academic Press, New York.

Cracraft, J. 1974. Phylogenetic models and classification. *Syst. Zool.* 23:71–90.

Cracraft, J. 1976. Covariation patterns in the postcranial skeleton of moas (Aves, Dinorthidae): A factor analytic study. *Paleobiology* 2:166–73.

Croizat, L., Nelson, G., and Rosen, D. E. 1974. Centers of origin and related concepts. *Syst. Zool.* 23:265–87.

Crowson, R. A. 1970. *Classification and Biology.* Aldine Publishing Co., Chicago.

Dall, W. H. 1903. Contributions to the Tertiary fauna of Florida, Part VI. *Trans. Wagner Free Inst. Sci.* 3:1219–1654.

Dennison, J. M., and Hay, W. W. 1967. Estimating the needed sampling area for subaquatic ecologic studies. *J. Paleontol.* 41:706–8.

Dodson, P. 1976. Quantitative aspects of relative growth and sexual dimorphism in *Protoceratops. J. Paleontol.* 50:929–40.

Durham, J. W. 1967. Presidential address: The incompleteness of our knowledge of the fossil record. *J. Paleontol.* 41:559–65.

Eldredge, N. 1968. Convergence between two Pennsylvanian gastropod species: A multivariate mathematical approach. *J. Paleontol.* 42:186–96.

Eldredge, N. 1971. The allopatric model and phylogeny in Paleozoic invertebrates. *Evolution* 25:156–67.

Eldredge, N. 1972. Systematics and evolution of *Phacops rana* (Green, 1832) and *Phacops iowensis* Delo, 1935 (Trilobita) from the Middle Devonian of North America. *Bull. Amer. Mus. Nat. Hist.* 147:45–114.

Eldredge, N. 1973. Systematics of Lower and Lower Middle Devonian species of the trilobite *Phacops* Emmrich in North America. *Bull. Amer. Mus. Nat. Hist.* 151:285–338.

Eldredge, N. 1974. Stability, diversity, and speciation in Paleozoic epeiric seas. *J. Paleontol.* 48:540–48.

Eldredge, N. and Gould, S. J. 1972. Punctuated equilibria: An alternative to phyletic gradualism. In T. J. M. Schopf (ed.), *Models in Paleobiology,* pp. 82–115. Freeman, Cooper and Co., San Francisco.

Eldredge, N., and Gould, S. J. 1977. Evolutionary models and biostratigraphic strategies. In E. G. Kauffman and J. E. Hazel (eds.), *Concepts and Methods of Biostratigraphy,* pp. 25–40. Dowden, Hutchinson, and Ross, Stroudsburg, Pa.

Eldredge, N., and Tattersall, I. 1975. Evolutionary models, phylogenetic reconstruction, and another look at hominid phylogeny. In F. S. Szalay (ed.), *Approaches to Primate Paleobiology (Contributions to Primatology,* Vol. 5, pp. 218–42). Karger, Basel.

Engelmann, G. F., and Wiley, E. O. 1977. The place of ancestor–descendant relationships in phylogeny reconstruction. *Syst. Zool.* 26:1–11.

Farris, J. S. 1976. Phylogenetic classification of fossils with Recent species. *Syst. Zool.* 25:271–82.

Farris, J. S. 1977. Review of G. F. Estabrook (ed.), "Proceedings of the Eighth International Conference on Numerical Taxonomy." *Syst. Zool.* 26:228–30.

Fischer, A. G. 1977. Biological events near the Cretaceous–Tertiary boundary: The deep ocean. *J. Paleontol. (Suppl.)* 51:10.

Flessa, K. W., and Imbrie, J. 1973. Evolutionary pulsations: Evidence from Phanerozoic diversity patterns. In D. H. Tarling and S. K. Runcorn (eds.), *Implications of Continental Drift to the Earth Sciences,* Vol. 1, pp. 247–85. Academic Press, New York.

Frazzetta, T. H. 1970. From hopeful monsters to bolyerine snakes? *Amer. Nat.* 104:55–72.

Frazzetta, T. H. 1975. *Complex Adaptations in Evolving Populations.* Sinauer Associates, Sunderland, Mass.

Gingerich, P. D. 1976. Paleontology and phylogeny: Patterns of evolution at the species level in Early Tertiary mammals. *Amer. J. Sci.* 276:1–28.

Gingerich, P. D. 1977. Stratophenetic approach to phylogeny reconstruction in vertebrate paleontology. *J. Paleontol. (Suppl.)* 51:12.

Gould, S. J. 1969. An evolutionary microcosm: Pleistocene and Recent history of the land snail *P. (Poecilozonites)* in Bermuda. *Bull. Mus. Comp. Zool.* 138:407–532.

Gould, S. J. 1973. Systematic pluralism and the uses of history. *Syst. Zool.* 22:322–24.

Gould, S. J. 1975a. Stochastic simulation of evolutionary history. In G. F. Estabrook (ed.), *Proceedings of the Eighth International Conference of Numerical Taxonomy,* pp. 72–75. W. H. Freeman and Co., San Francisco.

Gould, S. J. 1975b. Discussion remarks. In G. F. Estabrook (ed.), *Proceedings of the Eighth International Conference on Numerical Taxonomy*. W. H. Freeman and Co., San Francisco.

Gould, S. J. 1977. *Ontogeny and Phylogeny*. Belknap/Harvard University Press, Cambridge.

Gould, S. J., and Eldredge, N. 1977. Punctuated equilibria: The tempo and mode of evolution reconsidered. *Paleobiology* 3:115–51.

Gould, S. J., Raup, D. M., Sepkoski, J. J., Jr., Schopf, T. J. M., and Simberloff, D. S. 1977. The shape of evolution: A comparison of real and random clades. *Paleobiology* 3:23–40.

Griffiths, G. C. D. 1973. Some fundamental problems in biological classification. *Syst. Zool.* 22:338–43.

Hallam, A. 1973. *A Revolution in the Earth Sciences*. Clarendon Press, Oxford.

Harper, C. W., Jr. 1976. Phylogenetic inference in paleontology. *J. Paleontol.* 50:180–93.

Hazel, J. E. 1969. Faunal evidence for an unconformity between the Paleocene Brightseat and Aquia Formations (Maryland and Virginia). *U.S. Geological Survey Professional Paper* 650-C, pp. C58–65.

Hazel, J. E. 1977. Use of certain multivariate and other techniques in assemblage zonal biostratigraphy; examples utilizing Cambrian, Cretaceous, and Tertiary benthic invertebrates. In E. G. Kauffman and J. E. Hazel (eds.), *Concepts and Methods of Biostratigraphy*, pp. 187–212. Dowden, Hutchinson, and Ross, Stroudsburg, Pa.

Hennig, W. 1966. *Phylogenetic Systematics*. University of Illinois Press, Urbana.

Jackson, J. B. C. 1973. The ecology of molluscs of *Thalassia* communities, Jamaica, West Indies. I. Distribution, environmental physiology, and ecology of common shallow-water species. *Bull. Mar. Sci.* 23:313–50.

Jackson, J. B. C. 1974. Biogeographic consequences of eurytopy and stenotopy among marine bivalves and their evolutionary significance. *Amer. Nat.* 108:541–60.

Jackson, J. B. C. 1977. Some relationships between habitat and biostratigraphic potential of marine benthos. In E. G. Kauffman and J. E. Hazel (eds.), *Concepts and Methods of Biostratigraphy*, pp. 65–72. Dowden, Hutchinson, and Ross, Stroudsburg, Pa.

Jeletzky, J. A. 1965. Late Upper Jurassic and Early Lower Cretaceous fossil zones of the Canadian Western Cordillera, British Columbia. Geological Survey of Canada Bulletin 103, pp. 1–70.

Kauffman, E. G., and Bretsky, S. S. Systematics and ecology of lucinacean bivalves from the Brightseat and Aquia Formations (Paleocene) of Maryland and Virginia. In preparation.

Kauffman, E. G., and Buddenhagen, C. H. 1969. Protandric sexual dimorphism in Paleocene *Astarte* (Bivalvia) of Maryland. In G. E. G. Westermann (ed.), *Sexual Dimorphism in Fossil Metazoa and Taxonomic Implications; International Union of Geological Sciences, Ser. A*, No. 1, pp. 76–93. E. Schweizerbart'sche Verlagsbuchhandlung, Stuttgart.

Kauffman, E. G., and Hazel, J. E., eds. 1977. *Concepts and Methods of Biostratigraphy*. Dowden, Hutchinson, and Ross, Stroudsburg, Pa.

Keen, A. M. 1971. *Sea Shells of Tropical West America*, 2nd ed. Stanford University Press, Stanford, Calif.

Kennedy, W. J., and Cobban, W. A. 1977. The role of ammonites in biostratigraphy. In E. G. Kauffman and J. E. Hazel (eds.), *Concepts and Methods of Biostratigraphy*, pp. 309–20. Dowden, Hutchinson, and Ross, Stroudsburg, Pa.

Kier, P. M. 1977. The poor fossil record of the regular echinoid. *Paleobiology* 3:168–74.

Kuhn, T. S. 1962. *The Structure of Scientific Revolutions*. University of Chicago Press, Chicago.

Levinton, J. S. 1976. Review of "Evolution and Extinction Rate Controls" by A. J. Boucot. *Amer. Sci.* 64:84.

Maslin, T. P. 1952. Morphological criteria of phyletic relationships. *Syst. Zool.* 1:49–70.

Mayr, E. 1963. *Animal Species and Evolution*. Belknap/Harvard University Press, Cambridge, Mass.

Mayr, E. 1965. Numerical phenetics and taxonomic theory. *Syst. Zool.* 14:73–97.

Mayr, E. 1969. *Principles of Systematic Zoology*. McGraw-Hill, New York.

Mayr, E. 1976. *Evolution and the Diversity of Life*. Belknap/Harvard University Press, Cambridge, Mass.

McKenna, M. C. 1973. Sweepstakes, filters, corridors, Noah's arks, and beached Viking funeral ships in palaeogeography. In D. H. Tarling and S. K. Runcorn (eds.), *Implications of Continental Drift to the Earth Sciences*, Vol. 1, pp. 295–308. Academic Press, New York.

Menzies, R. J., George, R. Y., and Rowe, G. T. 1973. *Abyssal Environment and Ecology of the World Oceans*. John Wiley and Sons, New York.

Moore, R. C. 1948. Stratigraphical paleontology. *Bull. Geol. Soc. Amer.* 59:301–26.

Morgan, E. 1972. *The Descent of Woman*. Stein and Day, New York.

Nelson, G. 1973. Classification as an expression of phylogenetic relationships. *Syst. Zool.* 22:344–59.

Newell, N. D., and Boyd, D. W. 1975. Parallel evolution in early trigoniacean bivalves. *Bull. Amer. Mus. Nat. Hist.* 154:53–162.

Oaks, R. Q., Jr., and DuBar, J. R., eds. 1974. *Post-Miocene stratigraphy, central and southern Atlantic Coastal Plain*. Utah State University Press, Logan, Utah.

Olsson, A. A. 1964. *Neogene Mollusks from Northwestern Ecuador*. Paleontological Research Institution, Ithaca, N.Y.

Platnick, N. I. 1976. Drifting spiders or continents?—Vicariance biogeography of the spider family Laroniinae (Araneae: Gnaphosidae). *Syst. Zool.* 25:101–09.

Raup, D. M. 1972. Taxonomic diversity during the Phanerozoic. *Science* 177:1065–71.

Raup, D. M. 1976. Species diversity in the Phanerozoic: An interpretation. *Paleobiology* 2:289–97.

Rosen, D. E. 1974. Review of "Space, Time, Form. The Biological Synthesis" by Leon Croizat. *Syst. Zool.* 23:288–90.

Ross, H. H. 1974. *Biological Systematics*. Addison-Wesley, Reading, Mass.
Roughgarden, J. 1973. Possibilities for paleontology. *Science* 179:1225.
Schaeffer, B., Hecht, M. K., and Eldredge, N. 1972. Phylogeny and paleontology. *Evol. Biol.* 6:30–46.
Scheltema, R. S. 1977. Dispersal of marine invertebrate organisms: Paleobiogeographic and biostratigraphic implications. In E. G. Kauffman and J. E. Hazel (eds.), *Concepts and Methods of Biostratigraphy*, pp. 73–108. Dowden, Hutchinson, and Ross, Stroudsburg, Pa.
Schopf, T. J. M., Raup, D. M., Gould, S. J., and Simberloff, D. S. 1975. Genomic versus morphologic rates of evolution: Influence of morphologic complexity. *Paleobiology* 1:63–70.
Simpson, G. G. 1949. *The Meaning of Evolution*. Yale University Press, New Haven, Conn.
Simpson, G. G. 1953a. *The Major Features of Evolution*. Columbia University Press, New York.
Simpson, G. G. 1953b. *Evolution and Geography*. Condon Lectures, Oregon State System of Higher Education, Eugene, Oregon, December 1953, pp. 1–63.
Simpson, G. G. 1961. *Principles of Animal Taxonomy*. Columbia University Press, New York.
Simpson, G. G. 1969. *Biology and Man*. Harcourt, Brace, and World, New York.
Simpson, G. G. 1976. The compleat paleontologist? *Ann. Rev. Earth Planet. Sci.* 4:1–13.
Simpson, G. G., Roe, A., and Lewontin, R. C. 1960. *Quantitative Zoology*. Harcourt, Brace and Co., New York.
Sneath, P. H. A., and Sokal, R. R. 1973. *Numerical Taxonomy*. W. H. Freeman and Co., San Francisco.
Sokal, R. R., and Sneath, P. H. A. 1963. *Principles of Numerical Taxonomy*. W. H. Freeman and Co., San Francisco.
Sprinkle, J. 1976. Classification and phylogeny of "pelmatazoan" echinoderms. *Syst. Zool.* 25:83–91.
Stanley, S. M. 1970. Relation of shell form to life habits of the Bivalvia (Mollusca). *Geol. Soc. Amer. Mem.* 125.
Stanley, S. M. 1975. A theory of evolution above the species level. *Proc. Nat. Acad. Sci. USA* 72:646–50.
Stanley, S. M. 1977. Influence of rates of speciation and extinction on the diversity and evolutionary stability of higher taxa. *J. Paleontol. (Suppl.)* 51:26–27.
Stenzel, H. B. 1949. Successional speciation in paleontology: The case of the oysters of the *sellaeformis* stock. *Evolution* 3:34–50.
Sullivan, W. 1977. Sea-level canal could imperil marine life at either end, scientists say. *The New York Times*, October 10, 1977, p. 3.
Sylvester-Bradley, P. C. 1977. Biostratigraphical tests of evolutionary theory. In E. G. Kauffman and J. E. Hazel (eds.), *Concepts and Methods of Biostratigraphy*, pp. 41–63. Dowden, Hutchinson, and Ross, Stroudsburg, Pa.
Takeuchi, H., Uyeda, S., and Kanamori, H. 1970. *Debate about the Earth*, rev. ed. Freeman, Cooper and Co., San Francisco.

Tattersall, I. and Eldredge, N. 1977. Fact, theory, and fantasy in human paleontology. *Amer. Sci.* 65:204–11.

Taylor, J. D. 1968. Coral reef and associated invertebrate communities (mostly molluscan) around Mahe, Seychelles. *Phil. Trans. Roy. Soc. London Ser. B* 254:129–206.

Taylor, M. E. 1977. Late Cambrian of western North America: Trilobite biofacies, environmental significance, and biostratigraphic implications. In E. G. Kauffman and J. E. Hazel (eds.), *Concepts and Methods of Biostratigraphy*, pp. 397–425. Dowden, Hutchinson, and Ross, Stroudsburg, Pa.

Thompson, I. 1977. Biotic complexity in pelagic oceanic ecosystems. *J. Paleontol. (Suppl.)* 51:29.

Trueman, A. E. 1922. The use of *Gryphaea* in the correlation of the Lower Lias. *Geol. Mag.* 59:256–68.

Valentine, J. W. 1969. Patterns of taxonomic and ecological structure of the shelf benthos during Phanerozoic time. *Paleontology* 12:684–709.

Valentine, J. W. 1970. How many marine invertebrate fossil species?—A new approximation. *J. Paleontol.* 44:410–15.

Valentine, J. W. 1977. The geological record. In Th. Dobzhansky, F. J. Ayala, G. L. Stebbins, and J. W. Valentine (eds.), *Evolution*, pp. 314–48. W. H. Freeman and Co., San Francisco.

Valentine, J. W., and Campbell, C. A. 1975. Genetic regulation and the fossil record. *Amer. Sci.* 63:673–80.

Valentine, J. W., and Moores, E. M. 1970. Plate-tectonic regulation of faunal diversity and sea level: A model. *Nature (London)* 228:657–59.

# Cladism and Common Sense

## Niles Eldredge

The recent controversy concerning the proper approach to phylogenetic analysis in general, and in paleontology in particular, has basically been salutary. It signals a reawakening of the natural action of paleontologists to attempt to reconstruct the history of life. It also betokens a willingness to come to grips with fundamental methodological issues, with a view toward eventual improvement in methodology. At the very least, we should all be expected to understand thoroughly what it is that we do, why we do it, and how we actually go about doing it.

But the debates, discussions, and controversies that have emanated from this ferment suffer from the same sorts of ills that all other human enquiries tend to develop. The dialogue has been marred at times by shouting (as a substitute for careful thought), confusion of exposition (with consequent misunderstanding on the part of opponents), and a tendency to characterize the opposition in monolithic terms. Thus there has been as much tilting at windmills as reasoned discourse. My aim in this paper is to strip the problem of phylogenetic reconstruction methodology to its barest bones and then carefully recloak it in order to see if the genuine issues (which will still be subject to legitimate debate) cannot be separated from the needless misunderstanding and side issues (which perhaps we can get beyond in the near future).

Recent discussions of the value of one or another approach to phylogenetic reconstruction have all, by and large, failed to recognize that there are, minimally, three levels of complexity in such analysis. Adherents to phylogenetic systematics ("cladists"),[1] for example, have complained about the "ancestor worship" exemplified in the trees of many of their colleagues, and they in turn have been accused of leaving the *sine qua non* of phylogenies, i.e., ancestor–descendant relationships, out of their diagrams of phylogenetic relationship. Very simply, few people thus far have bothered to make explicit the fundamental distinction between a cladogram and a tree (Nelson, in preparation).

Cladists (e.g., Schaeffer, Hecht, and Eldredge, 1972) have sometimes claimed that although cladograms are not "evolutionary event-o-grams," they nonetheless have all the basic attributes of trees, and are as good since they contain all that can be wrought from a phylogenetic analysis. For their part, opponents to cladism (e.g., Campbell, 1975) have acted as though cladograms and trees are identical in that cladograms specify actual evolutionary events and ancestor–descendant relationships but that cladists are capricious in refusing to identify a particular, known taxon as that elusive "hypothetical common ancestor." It is this failure to distinguish fully between cladograms and trees that has caused most of the confusion so far. As will be developed below, cladograms are simpler and more general statements than phylogenetic trees and are especially to be distinguished inasmuch as trees specify the nature of the phylogenetic relationship between any two taxa (even if one is imaginary), whereas cladograms do not.

Another source of confusion in the debate is the predilection of some authors for rationalizing their preferred methodology by reference to some specific body of evolutionary theory. Thus Brundin (1972) and Eldredge and Tattersall (1975) not only claimed congruence between current models of animal speciation and phylogenetic systematics but went on to claim that the congruence implied that phylogenetic systematics (i.e., cladism) reflects the underlying nature of the evolutionary process more accurately and is therefore the method of choice for the reconstruction of phylogenetic history. On the other side of the coin, Gingerich's stratophenetic approach (e.g., 1976a, pp. 15–16) is a methodological outgrowth of his adherence to phyletic gradualism. It now appears to me more suitable to view phylogenetic reconstruc-

[1] "Phylogenetic systematics" is the translation of Hennig's perhaps overly appropriative term. "Cladism" and, still worse, "Hennigism" (or "Hennigianism") are inaccurate if not downright offensive. I'll stick to "cladism" because it is shorter, and for want of a better term.

tion as a wholly separate corpus of methodological theory, replete with its own axioms, assumptions, and propositions, which can be developed logically and consistently starting merely with the initial grand assumption that life *has* a phylogenetic history. All we seek is a methodological strategy to recapture, as nearly as possible, that history. Refined methodological formulations based on some specific notion of how the evolutionary process works add the obvious danger that these notions may well be wrong.

In retrospect, publication of Hennig's (1966) book seems to have contributed one major, fundamental insight into contemporary systematics. This insight has finally made explicit, after all these years, what it is that allows us to do systematics in the first place. For the most part, we had all thought of two basic kinds of similarity among organisms: "true" evolutionary similarity, i.e., resemblance inherited from a common ancestor, and convergence and parallelism, i.e., "false," or misleading, adaptive similarity. Hennig (1966) merely made explicit what, in retrospect, seems painfully obvious: There are, at any given level of analysis, two kinds of evolutionary similarities. In any instance, when comparing any two taxa, we may be dealing with (1) shared evolutionary novelties inherited from an immediate common ancestor and thus not to be found in any other taxon, (2) resemblances inherited from some more remote common ancestor such that one or more other taxa descended from the same ancestor may retain the same similarity, or (3) similarities due to parallelism and convergence. Only the first kind of similarity is valid evidence of close affinity between the two taxa. Figure 1 illustrates the distinction between the two kinds of evolutionary similarities. It is noteworthy that most of the prominent critics of phylogenetic systematics (e.g., Mayr, 1974, p. 98; Simpson, 1975, p. 14) have recognized the essential validity of this distinction.

Thus the basic problem in phylogenetic analysis is to determine as accurately as possible what the patterns of shared evolutionary novelties are among the taxa under study. The methods that bear on this problem have been discussed (with varying degrees of clarity and accuracy) by Hennig and many of his followers but are also found scattered in the literature of systematics (e.g., chapter 1 of Clark, 1959), extending at least as far back as the preceding century. The results of such an analysis are presented on one (or, if conflict persists, more than one) cladogram, which is *a branching diagram depicting the pattern of shared similarities thought to be evolutionary novelties ("synapomorphies") among a series of taxa.* With the addition of certain assumptions,

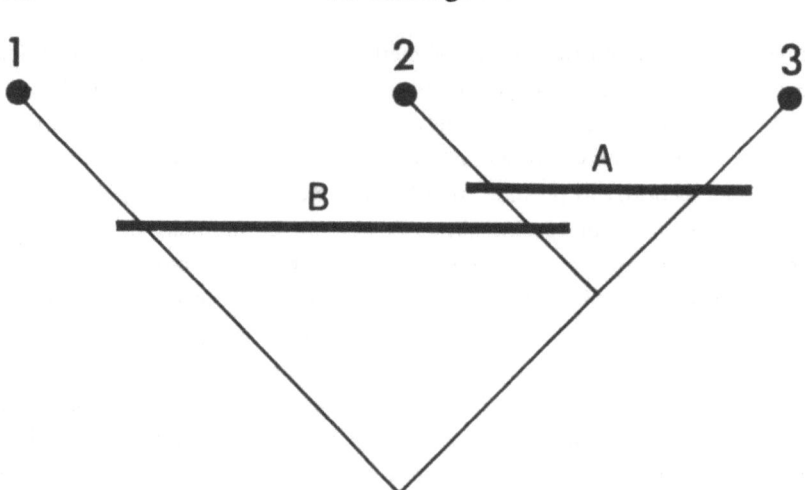

Figure 1. A cladogram of three taxa. Taxa 2 and 3 share an evolutionary novelty (a "synapomorphy," indicated by the black bar labeled A) not possessed by taxon 1. Taxa 1 and 2 share a primitive similarity (a "symplesiomorphy," indicated by the black bar labeled B) not present in taxon 3. Shared possession of a derived similarity by taxa 2 and 3 is direct evidence that they are more closely related to each other than either is to taxon 1. Shared possession of a primitive state by taxa 1 and 2, in contrast, is not indicative of any special relationship between them.

and upon analysis of additional information, we can make more precise statements concerning actual evolutionary relationships among taxa. Thus I define a phylogenetic tree as *a diagram* (not necessarily branching!) *depicting the actual pattern of ancestry and descent among a series of taxa*. Finally, a phylogenetic scenario is *a phylogenetic tree with an overlay of adaptational narrative*.

I contend that these three levels of phylogenetic analysis are all valid scientific activities, but they are linked along a scale of relative complexity such that the analysis used to produce a cladogram must precede the construction of a tree (even if a cladogram is not actually drawn by the investigator), and there must be a tree before a scenario can be constructed. Thus, properly speaking, there should be no debate over whether we should construct cladograms *instead* of trees, or vice versa. Rather, the only valid points to be debated are (1) the procedures for cladogram, tree, and scenario construction and (2) whether or not the additional assumptions we must make when stepping to the next, more complex level of analysis are worth it. Thus, contrary to some of my earlier opinions (e.g., Schaeffer, Hecht, and Eldredge, 1972;

Eldredge and Tattersall, 1975), I no longer oppose the construction of phylogenetic trees outright, or for that matter, scenarios (which are, after all, the most fun), but merely point out that, in moving through the more complex levels, we inevitably become further removed from the original data base in adding assumptions and *ad hoc* (and largely untestable) hypotheses. As long as we understand precisely what we are doing at each step in the analysis, which includes having an adequate grasp of the probability that we are wrong and what the assumptions are that we have added along the way, there no longer seems to me any reason for anyone to tell anyone else what *not* to do.

# Cladograms

A cladogram is nothing more than a branching diagram depicting the nested pattern of synapomorphies (or "shared derived characters") among the taxa under study. A character state considered derived at one level becomes primitive when finer relationships within a group are sought. Mammalian hair is a synapomorphy for Mammalia but, as such, is primitive for the group—we cannot recognize finer levels of relationship among mammals on the basis of "presence of hair," even though some Recent mammals are hairless. Hennig (1965, p. 107) has referred to this necessarily nested, hierarchical arrangement of primitive and derived conditions as the "heterobathmy of synapomorphy," and Nelson (in preparation) has pointed out that symplesiomorphies (shared primitive states) are simply synapomorphies considered at the wrong (i.e., too refined) level. All character states are apomorphic (derived) *sometime* in the phylogenetic history of the group under study.

Thus the problem is to "discover" this nested pattern of synapomorphies we know must be there. There are two general approaches to this problem. Nelson (personal communication, 1976) and others have recently suggested that, inasmuch as we know such a pattern must exist, we simply map nested sets of resemblance among taxa and that this pattern is likely to be correct. Should more than one pattern emerge, any conflict can be resolved by reference to a parsimony argument: Elect the hypothesis that minimizes the number of evident parallelisms.[2] This hypothesis is susceptible to additional testing and potential refutation through consideration of additional data. In a sense, the cladogram "falls out" of the analysis in much the same way as a phenogram is the direct output of some computer program in numerical taxonomy.

The more conventional approach to the production of caldograms involves explicit analysis of characters in terms of primitive and derived states. This procedure is conceptually similar to conventional evolutionary approaches. The only real difference is the explicit addition of the distinction between primitive and derived resemblance at any given level (degree) of relationship. Thus, in this approach, cladograms are constructed after the systematist has analyzed the known states of all (presumably) homologous characters which display some amount of among-taxa variation.

Several general approaches to the analysis of primitive versus derived character states have been advocated. All approaches stem from nineteenth-century practice. There are no easy ways.[3] Nor are there any "sure-fire" ways.

The most commonly cited approach, the "commonality principle" (see Hecht, 1976; Kluge and Farris, 1969; Schaeffer, Hecht, and Eldredge, 1972), is actually an *a priori,* explicit statement of the "recognition" technique of cladogram construction already mentioned. The argument is, simply, that a character state widespread within a group is likely to be primitive for that group. It may in addition be synapomorphic for the entire group or primitive for a much more inclusive taxon. Thus, in assessing primitive versus derived states, it is further useful to compare the group under study

---

[2] If, as is nearly always the case when more than one character is examined, more than one possible pattern of synapomorphy is evident, it is clear that only one can be "correct." Other apparent patterns of synapomorphy are therefore false, and the similarities upon which they are based are, by definition, parallelisms (see also Platnick, 1977). In the case where the investigator is making conscious evaluations as to which character states are primitive and which derived, conflicting cladograms imply actual analytic error, as well as the fact of parallelism. As a result of the relatively recent production of a plethora of cladograms, parallelism turns out to be a far more common evolutionary phenomenon than even most of its more ardent aficionados had thought.

[3] One of the main complaints about cladism seems to be that there is no pat, simple method for recognizing apomorphic versus plesiomorphic conditions. But this begs the question once the (seemingly undeniable) proposition is accepted that such distinctions lie at the heart of phylogenetic analysis, as all involved in phylogenetic research now seem to agree. Just because the job now seems more demanding is hardly grounds for rejection of a method.

with its closest relative (the "sister taxon").[4] Should the character state be found to be widespread within the sister taxon as well, we would hypothesize it to be primitive for the entire sister group and thus, of course, primitive for the group under immediate study. Thus derived states are those with limited distributions, which results in the patterns of special resemblance discussed above.

Another approach to the problem involves morphogenetic sequences. Although we raise our eyebrows at the literal reading of ontogeny as phylogeny, especially where early ontogenetic stages are held to represent, or to *be*, ancestral adult stages, it is nonetheless true that comparison of (presumably) homologous characteristics of progressively earlier developmental stages of (presumably) phylogenetically related taxa reveals an increasingly general pattern of resemblance among them. We all know that the embryos of representatives of the various vertebrate classes exhibit more obvious characters in common than do the adult stages. Thus ontogenetic sequences are of direct benefit in the elucidation of patterns of relationship. Explicitly, we may hypothesize a character as primitive (in an adult stage) if we see that other adults without that character state nevertheless possess a similar character state in their development (*pace* fans of neoteny).

Thus the problem boils down to mapping the relative ubiquity of character states among all stages of life history of all organisms under study (and their nearest relatives if possible), and deriving, whether by explicit *a priori* argument or by "letting the chips fall where they may," one or more cladograms based on shared possession of special resemblances which we hypothesize to be evolutionary novelties. This is how all intrinsic properties of organisms (i.e., everything from molecules to behavior) are to be evaluated in a comparative, phylogenetic sense.

---

[4] Critics (e.g., Campbell, 1975, p. 89) complain that such a procedure is circular, i.e., that we must know a higher-level phylogeny before we can analyze a lower-level one, yet the method seems to advocate a pair-wise procedure whereby cladograms are constructed by recognition of sister groups among the basal elements (taxa) under study (usually species). It is legitimate, however, to formulate and test alternative hypotheses of relationship within the context of a larger hypothesis which is accepted for the sake of the analysis. How could it be otherwise? In a more conventional context, of the three currently considered possible candidates as the ancestor of *Archaeopteryx*, all are diapsid "reptiles." No one would bother to search for a direct ancestor among the nondiapsid Amniota, still less among the nonamniote Tetrapoda, or the nonvertebrate Chordata. We must assume we have gotten somewhere during the last few hundred years of research in systematics.

Data pertaining to the distribution of organisms in space and time are also, at least in part, a product of the phylogenetic histories of the organisms. Thus there is a prevailing tendency among paleontologists to use stratigraphic evidence in the formulation of phylogenetic trees. It is precisely in the construction of trees, not cladograms, that such information is relevant, as I shall discuss more fully below. Insofar as construction of cladograms is concerned, one point of view (e.g., Schaeffer, Hecht, and Eldredge, 1972) maintains that distributional data, although certainly suggestive, tend to add noise to the evolutionary signal found in the intrinsic data, which are almost solely the product of phylogenetic history. Others (e.g., Bretsky and Gingerich, both in this volume) maintain that all sorts of data must be brought to bear on a given problem. The point becomes moot if a cladogram is regarded strictly as a nested pattern of synapomorphies and its construction a matter of recognizing that pattern. It is, properly speaking, only when we are building trees (and scenarios) that distributional data become at all relevant.

One final aspect of cladograms deserves attention before we pass to phylogenetic trees; that is, rarely will an analysis produce only one cladogram for a group of taxa. We usually get a series of conflicting hypotheses. The task then becomes to reject all but the least unlikely hypothesis (cladogram). Inasmuch as conflicting cladograms imply false recognition of synapomorphies (i.e., parallelisms), the most straightforward approach would appear to be simply to pick that cladogram which minimizes the number of necessary parallelisms, so-called parsimony. There are, minimally, two potential problems with strict adherence to parsimony in these situations. First, it is conceivable that the cladogram requiring the fewest parallelisms might be based on symplesiomorphies rather than synapomorphies. In such a situation, one must compare the group under study with the "out group" (i.e., putative sister group) to formulate an explicit understanding of primitive versus derived within the group (G. F. Engelmann, personal communication, 1976). In response to this objection, it should be noted that the retention of plesiomorphies in such a highly ordered, patterned form is unlikely. That is to say, plesiomorphies should be expected to be retained in a quasi-random fashion among characters and not display the sort of "linkage" that would have to be present to lead one astray. This is a probability argument, however, and it is perhaps true that outgroup comparison, i.e., the explicit evaluation of primitive versus derived states for all characters, is a logical necessity in the case of cladogram conflict.

Another potential objection to the simple, straightforward use of parsimony is that it may cause us to take the data overly seriously. For instance, it has been argued that highly integrated functional complexes might be taken as a single megacharacter rather than, say, 20 to 50 separately enumerable characters. In such a situation, one's view of parsimony is likely to oscillate wildly as one grapples with various anatomical regions and views them as functional complexes. Hecht and Edwards (1976) and Hecht (1976) have ranked *kinds* of morphological features on a scale of one to five, according to their assessment of the relative likelihood of each of these kinds of features exhibiting parallelism. Thus they rate "loss characters" (e.g., blindness) as particularly subject to parallelism. A cladogram whose synapomorphic pattern is based on such loss characters is, in their view, less likely to be correct than one based on structures which are clearly evolutionary novelties. But the danger with such *a priori* notions of trustworthiness is that blindness, for instance, may be an excellent synapomorphy. Blindness cannot be used as a synapomorphy within the Class Trilobita as a whole but is an excellent synapomorphy for the Conocoryphidae, a family of Cambrian ptychoparioids. Probably the best way to test a series of conflicting cladograms is to gather additional data (i.e., other characters), which amounts to an ongoing application of the principle of parsimony. However risky raw parsimony may be, we are still better off if that is our primary criterion. The examples below evaluate conflicting cladograms from the point of view of parsimony, though additional arguments on the probability of parallelism among the different items of data causing the conflict are also made.

## An Example—The Acastacea

The cladogram in figure 2 depicts what seems to be the least objectionable hypothesis of distribution of synapomorphies among phacopid trilobites (Suborder Phacopina *sensu* Treatise). The three major groups (Infraorder Pha-

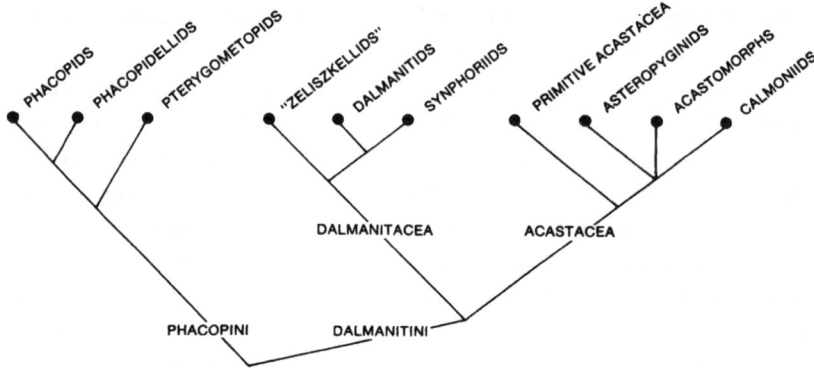

Figure 2. The cladogram depicting the preferred hypothesis of the distribution of synapomorphies among the major groups of the trilobite suborder Phacopina, as discussed in the text.

copini and Infraorder Dalmanitini with two superfamilies, Dalmanitacea and Acastacea) are defined and recognized primarily on patterns of the auxiliary impression system on the anterior glabellar lobe, though other characters are also used. Detailed arguments on the phylogeny of the Phacopina will be the subject of a separate paper, and I shall focus on the Acastacea to illustrate aspects of construction, and some peculiarities, of cladograms.

Eldredge (1971) reviewed patterns of expression of the auxiliary impression system and recognized a fundamental dichotomy between the ovate-circular pattern of the Phacopini (Pterygometopidae + Phacopidae) and the "triangular" patterns of the Dalmanitini. (See figure 3 for a comparative diagram of basic patterns of the auxiliary impression system among Phacopina.) Recent work (Clarkson, Eldredge, and Henry, 1977) has demonstrated a further connection between the Pterygometopidae and Phacopidae, thus adding corroboration to the hypothesis. However, as pointed out by J.-L. Henry (personal communication 1975), most, if not all, Ordovician zelliskellids and "acastids" (i.e., all Ordovician "dalmanitids" save pterygometopids) possess a lozenge or diamond-shaped pattern of the auxiliary impression system, with no median window devoid of scars. Siluro-Devonian Dalmanitidae and Synphoriidae possess a greatly simplified, truly triangular pattern. The remainder of the Siluro-Devonian "dalmanitids"—the acastomorphs, calmoniids, and asteropyginids—possess a highly distinctive triangular pattern consisting of two pairs of well-developed rows of impressions radiating from the posteromedial area of the anterior glabellar lobe. The widest portion of the field is at the anterior margin of the glabella, just proximal to the anterior border furrow or facial suture (whichever is nearest). The pattern is unmistakable and virtually invariant and serves as an excellent synapomorphy linking these three groups. Ordovician "acastids" (e.g., *Kloucekia*) retain the primitive, diffuse lozenge pattern, though a specimen of *Ormathops* sp. from the Ordovician of Czechoslovakia (in the collections of the Museum of Comparative Zoology, Harvard University) shows a triangular pattern similar to that of the Acastacea, save the absence of a median "window." Other characters have long served to link acastids with calmoniids, but the separation of the asteropyginids from the dalmanitids and their alliance with the acastids presents something of a departure from convention. However, several authors (e.g., Gandl, 1972; Struve, 1959, Shergold, 1966) have noted "asteropyginid tendencies" among certain acastomorphs, based on glabellar and pygidial similarities, and there is little doubt that the Acastacea as delineated in figure 2 is a monophyletic group and the sister group of the Dalmanitacea (in its new, restricted sense).

Figure 4 shows the trichotomous arrangement of the Acastacea. The syna-

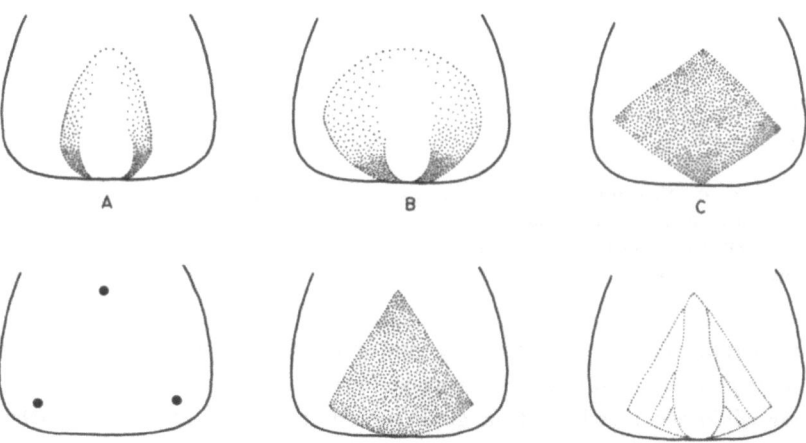

Figure 3. Major variation in patterns of expression of the auxiliary impression system on the anterior glabellar lobe of various Phacopina. Wholly diagrammatic. (*A*) Ovate pattern (Pterygometopidae, early Phacopacea). (*B*) Circular pattern (advanced phacopids). (*C*) Lozenge pattern (Ordovician Dalmanitacea). (*D*) Dalmanitid triangular (dalmanitids and synphoriids). (*E*) Primitive acastid triangular (Ordovician "acastids," note absence of median "window"). (*F*) Acastid triangular (Siluro-Devonian Acastacea, note median "window").

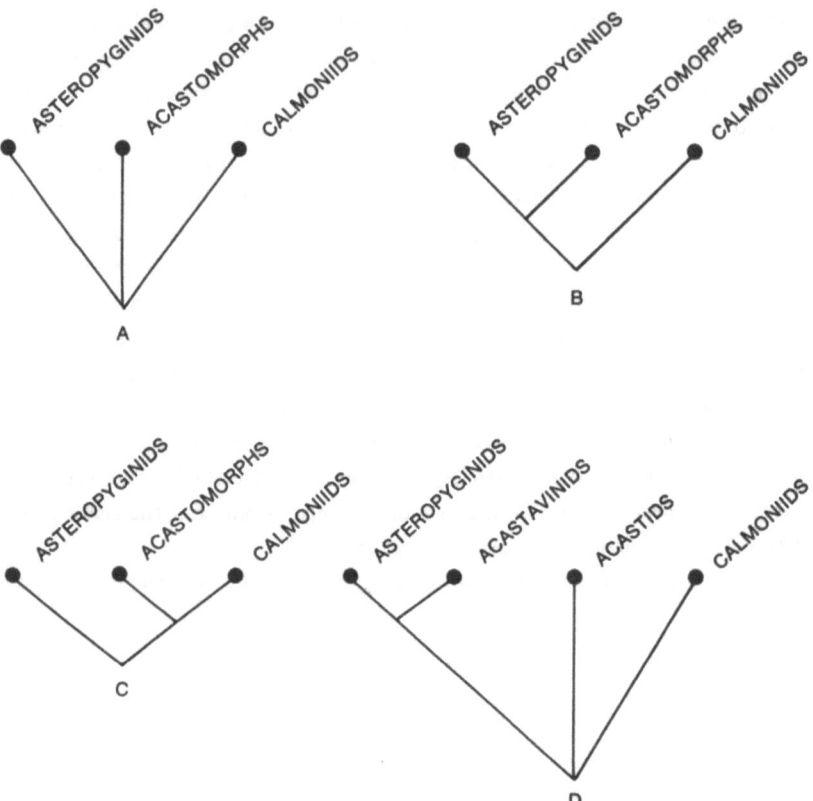

Figure 4. Alternative cladograms for relationships among *Acastacea*. (A) General trichotomous arrangement. (B and C) Two of the three dichotomous possibilities among the three taxa. (D) A different hypothesis, based on recognition of two groups of acastamorphs, as discussed in the text.

pomorphic pattern of the auxiliary impression system serves only to link the three groups and is therefore primitive for the group as a whole. That is to say, all known members of the group possess this pattern as defined. Further refinement of the analysis must rely either on other specific aspects of the pattern which may display variation within Acastacea (e.g., mode of expression—scars, calluses, etc., and presence or absence of subsidiary rows of impressions) or on entirely different features. Figures 4B and 4C depict two of three possible resolutions of the trichotomy, though I cannot suggest any synapomorphies which would lead to choosing one cladogram over the other.

The third possibility, i.e., that the calmoniids and asteropyginids comprise a sister group, is even harder to substantiate. This state of affairs arises because the acastomorphs seem so plesiomorphic in all respects and form a sort of base-level connection between the relatively (and quite differently) derived calmoniids and asteropyginids. Thus this primitive acastomorph group is probably not strictly monophyletic. Following the arrangements of Struve (1959) and others, we can isolate the acastavine group of acastomorphs (i.e., those showing "asteropyginid tendencies") and arrive at the cladogram of figure 4D. This the best that can be accomplished given the present state of our knowledge.

Finally, I present a cladogram of higher-level relationships among the calmoniids to illustrate the use of outgroup comparison and ontogeny in the elaboration of a series of cladograms and the resolution of manifest conflicts in hypothesized patterns of synapomorphy. Referring to figure 4D, we note that we must include the acastids *and* the acastavinids + asteropyginids as the sister group of the calmoniids, owing to lack of resolution of that particular trichotomy.

The calmoniids are a group of some 34 genera and subgenera of Acastacea endemic to the Malvino-Kaffric faunal province of the southern hemisphere (Eldredge and Ormiston, 1979). With the exception of one very primitive, acastid-looking genus in the Silurian, all are Upper Lower to Lower Middle Devonian in age. With the exception of four or five generally plesiomorphic genera retaining clear acastid morphology (e.g., *Phacopina,* "*Acastoides*"), the taxa represent a morphologically highly diverse group, which makes definition of the entire group by one or more synapomorphies difficult. The calmoniids do seem to be linked together and set off from the other Acastacea by presence of truly small eyes relative to total cephalon size, absence of a well-developed palpebral lobe demarcated from the palpebral area by a distinct furrow, and relatively light impression of the axial furrows. Should one or more of these features be primitive for Acastacea, then at least the opposite states as developed in the rest of the Acastacea would be considered synapomorphies linking *them* and separating them from the calmoniids. Should all these features prove primitive for Acastacea, then of course we still have no evidence that the calmoniids are monophyletic.

For the remainder of this discussion, I shall assume that the calmoniids constitute a coherent, monophyletic group. Figure 5 illustrates the pattern of synapomorphies among the four major groups of calmoniids recognized by

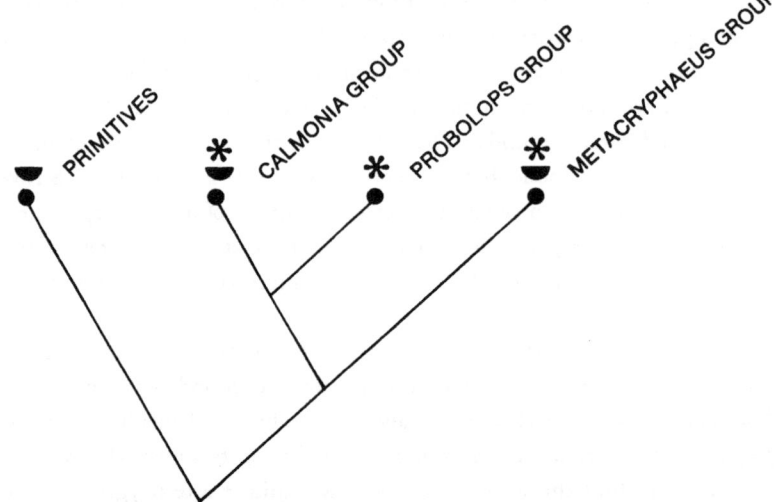

Figure 5. A cladogram depicting the preferred hypothesis of relationships within the calmoniids. Groups containing one or more genera with posteriorly rounded, micropygous pygidia are designated with a semicircle; groups with one or more genera with five pairs of marginal pygidial spines are designated with a star.

Eldredge and Branisa (unpublished manuscript). The primitive group is not considered monophyletic but is lumped together for convenience. The *Calmonia* group is allied by a number of cephalic and pygidial features, especially cephalic features distal to the facial suture. The *Probolops* group shares many of these features but is further distinguished by the development of a series of "metafixigenal" spines serially homologous on the palpebrum, fixigena, thorax, and pygidium. The *Metacryphaeus* group is morphologically the most diverse (including, *inter alia,* the nominate genus, which is the most dalmanitidlike of all calmoniids, and *Bouleia,* which is a *Phacops* homeomorph). Glabellar furrows tend to be deepened and rearranged in this group, and the group as a whole is clearly linked to the *Calmonia–Probolops* group by the intermediate genus *Kozlowskiaspis* (Eldredge and Branisa, unpublished manuscript). A more detailed analysis of the relationship among all genera of calmoniids currently recognized (Eldredge and Ormiston, 1979) is in preparation.

Recognition of four major groups of calmoniids and the pattern of relationships among them depicted in figure 5 is by no means the only arrangement suggested by the similarities among these genera. Two other patterns are shown by symbols in figure 5. Groups marked with a semicircle have at least one genus with a posteriorly rounded, micropygous pygidium. These include *Phacopina* and *"Acastoides"* within the primitive group, *Schizostylus* in the *Calmonia* group, and *Parabouleia* in the *Metacryphaeus* group. Posteriorly rounded, micropygous pygidia are known in some acastids, but the basic shape of the pygidium of members of the Acastacea is triangular, with a pointed terminus, and there is no doubt that this is the primitive condition for Acastacea (indeed, probably for all Dalmanitini and perhaps even for Phacopina in general) and is therefore primitive for the calmoniids as well. Furthermore, most calmoniids (by comparison of all genera within the group) have triangular pygidia. Thus the posteriorly rounded, micropygous condition of the pygidium is a derived feature within the calmoniids. We are thereby faced with a conflict in the distributions of evident synapomorphies within the Calmoniidae. Why not, then, recognize a group of calmoniids embracing the genera with this derived condition? This remains a distinct possibility. I have chosen not to do so primarily on the grounds of parsimony, since pygidial shape is essentially a "simple" character. In point of fact, the pygidia of these taxa have little else in common, aside from their shape. When compared with the number of other characters (admittedly, not fully spelled out and argued in this paper) that are used to define the groups in figure 5, it seems most parsimonious to postulate parallelism in pygidial shape. Furthermore, other acastids (noncalmoniids) occasionally exhibit this feature, and it is also—and more obviously—a parallelism. Finally, there is a functional consideration: Rounded pygidia enable a more perfect occlusion with the ventral surface of the anteriorly curved cephalic margin upon enrollment. Thus we might envision this adaptive complex arising independently in a number of different lineages. The problem with this kind of reasoning is that we think we understand the functional significance of a particular morphologic configuration and thus readily imagine its adaptive significance and therefore are automatically predisposed to concede the strong possibility (even probability!) that this configuration could have evolved more than once. But, in fact, such arguments beg the question (the adaptation could have arisen only once) and in the end, we only focus on those complexes whose functional significance we think we grasp and ignore the other complexes because their functional significance remains obscure to us.

Referring again to figure 5, note that the stars indicate groups which contain at least one genus with five pairs of marginal spines distributed around the pygidial margin. This is an excellent example of the bearing of ontogenetic information on the resolution of conflicts in apparent patterns of synapomorphy. The genera include *Calmonia* and *Pennaia* (*Calmonia* group), *Cryphaeoides* (*Probolops* group), and *Metacryphaeus* (*Metacrypaheus* group). Is this pattern primitive or derived within Calmoniidae? It is uncommon within Calmoniidae, and in fact, the primitive condition within Calmoniidae would appear to be a rather short, triangular, and posteriorly pointed pygidium with a smooth margin. When we examine the composite sister group of the Calmoniidae, however, we find a curious distribution of the condition of five pairs of spines around the pygidial periphery: The condition is consistently present in all asteropyginids (hence the name) and is tacitly considered to be a synapomorphy linking the component genera and defining the group (there are other and perhaps better synapomorphies on the cephalon). Some of the acastavinids (e.g., *Acastava*, *Acastella*) possess denticles (short lappets or spines) around the periphery (part of the "asteropyginid tendencies" noted by several authors cited above). However, Shergold (1966, p. 184) points out that in *Acastava atava*, the spines are developed only on the internal mold, i.e., are not visibly expressed on the external cuticular surface (see fig. 5). Richter and Richter (1952, p. 99) make a similar observation on *Acastoides henni henni*. Ontogeny enters the picture with Shergold's further observation (1966) that in young holaspid specimens of *Acastocephala macrops*, such denticulations are indeed present but are resorbed during ontogeny and are wholly absent in the large holaspid specimens. Finally, there is no evidence for marginal pygidial spines in any of the acastids in the restricted sense of figure 4D, except in *Acastoides henni*. Thus the distribution of these spines within Acastacea as a whole is spotty, and the character appears to be "latent" in some taxa and present only in the early stages of ontogeny in some other taxa, which suggests that presence in the adult stage is neotenic. Other, more remotely related groups of trilobites typically show resorption of cuticular spines in ontogeny (e.g., genal spines in *Phacops rana milleri*). Have we, then, grounds for concluding that this condition is primitive for Acastacea? Possibly so, but probably not. Certainly, expression of this feature in adults is not primitive for Acastacea (because most members lack spines and they are absent in Dalmanitacea, the sister group). But it remains conceivable that possession of such spines in the early stage of ontogeny is primitive for the group. Thus expression of this feature on the adult cuticle is most reasonably inferred as derived. Should we then unite all asteropyginids, certain acastavinids, and the four aforementioned calmoniid genera into a single group? Consideration of other possible synapomorphies and the principle of parsimony rules out this course, but why not at least consider the four calmoniid genera exhibiting this condition as constituting a distinct monophyletic group? Again we appeal to parsimony as the best way to resolve the conflict but further observe that the character is apparently latent, i.e., present early in ontogeny in at least some of the members of the sister group, and is thus probably primitive for Calmoniidae, though not as an adult feature. Yet its appearance in adults of other Acastacea might prompt us to propose its appearance by neoteny independently within Calmoniidae. Formally, by this reasoning alone, we still do not know whether it appeared as an adult feature only once (most parsimonious for this single character state) or two or three times. The only point here is that the apparent lability of the character arouses suspicion and helps us to reject the pattern as synapomorphic for the four calmoniid taxa. Similarly, Campbell (1977) has devised an intriguing functional explanation for the presence of such pygidial spines (in another context, and with a different group of Dalmanitini). I shall not recapitulate the functional argument here, since it parallels the functional argument already adduced concerning rounded pygidia and suffers from the same defects as applied to the resolution of conflicting patterns of apparent synapomorphies.

Finally, since synapomorphies are homologies, by definition (the converse is not true, since symplesiomorphies are also homologies), and pseudo-synapomorphies are either not homologies or homologies at the wrong level, it is relevant that the marginal pygidial spines of *Cryphaeoides* appear to be homologous with the metafixigenal series of *Tarijactinoides* and *Probolops* (two other genera of the *Probolops* group) which have "migrated" to the pygidial margin. Thus the condition "five pairs of marginal pygidial spines" may actually involve nonhomologous spines, at least in the case of *Cryphaeoides*.

In such a fashion, we arrive at the least objectionable cladogram for a series of organisms assumed at the outset to have some relationship with one another. The resultant cladogram is testable by reanalysis of parsimony arguments adduced to resolve conflicts in apparent patterns within existing data and by addition of new data (additional characters among the taxa involved in the hypothesis) but not new taxa, since this would merely change the hypothesis (E. S. Gaffney and N. Platnick, personal communications, 1977).

The resultant cladogram is purely a hypothesis of patterns of synapomorphy among the taxa considered. It is not a phylogeny in the usual sense of the word, but merely a pattern of morphological features.

## Phylogenetic Trees

A phylogeny is an explicit statement concerning the exact nature of the evolutionary relationship between two, or among three or more, taxa. The fundamental notion of phylogenesis is ancestry and descent. Thus the reconstruction of patterns of ancestry and descent is the goal of phylogenetic analysis. Such patterns are expressed in diagrams called phylogenetic trees.

Patterns of ancestry and descent have evolutionary meaning at the population and, more importantly, the species level. This follows from simple consideration of our theories of the evolutionary process; that is, individuals reproduce (i.e., show patterns of ancestry and descent among them) but do not evolve per se. Rather, it is taxa—populations and especially species—that do the evolving (Eldredge, 1979). Moreover, genera and taxa of even higher categorical rank are classificatory conventions, i.e., the taxa presumably are monophyletic clusters of subordinate taxa, including species. When one genus, say, is held to be ancestral to another, what is meant is that a species classified as genus A gave rise to a species classified as genus B. Were that ancestral species known, it would have to possess some special resemblance (i.e., a synapomorphy) with the species in genus B to cause us to suspect a special relationship (ancestry) with that species in genus B. And thus, to retain a strict monophyletic system, the ancestral species should be regarded as a member of genus B, not genus A. This amounts to saying that a genus or family cannot give rise to another genus or family, which of course is only consonant with what we think is going on in the evolutionary process. Thus the following discussion of the analysis of ancestry and descent among taxa and the construction of phylogenetic trees is, strictly speaking, limited to species. For those wishing to retain Simpson's (1961) broader definition of monophyly, however, most of the remarks hold equally well for taxa of any categorical rank.

There is a single, simple, general form of the phylogenetic tree for any two taxa suspected to have a special evolutionary relationship involving ancestry and descent (fig. 6A). This general tree—the "A-tree" ("A" for ancestor)—looks like a cladogram, with the major, and important, difference that a taxon is specified at the branching node. This general A-tree subsumes all five ancestor–descendant relationships possible between any two taxa. If for $A_i$, $i = 1$, then the two possible trees are as shown in figures 6B and 6C; if $i = 2$, then taxon 2 becomes the ancestor and the same two possible forms of the tree are repeated, with the ancestor–descendant relationship simply inverted. Finally, if $i \neq 1$ or 2, then we have a third taxon giving rise to both 1 and 2 (fig. 6F). We may specify that taxon ($i = 3$, for instance) or merely postulate its existence. In the latter case, we call it "X," corresponding to the "X-tree" of Nelson (in preparation).

These five possibilities (actually, three basic forms of a tree) correspond to

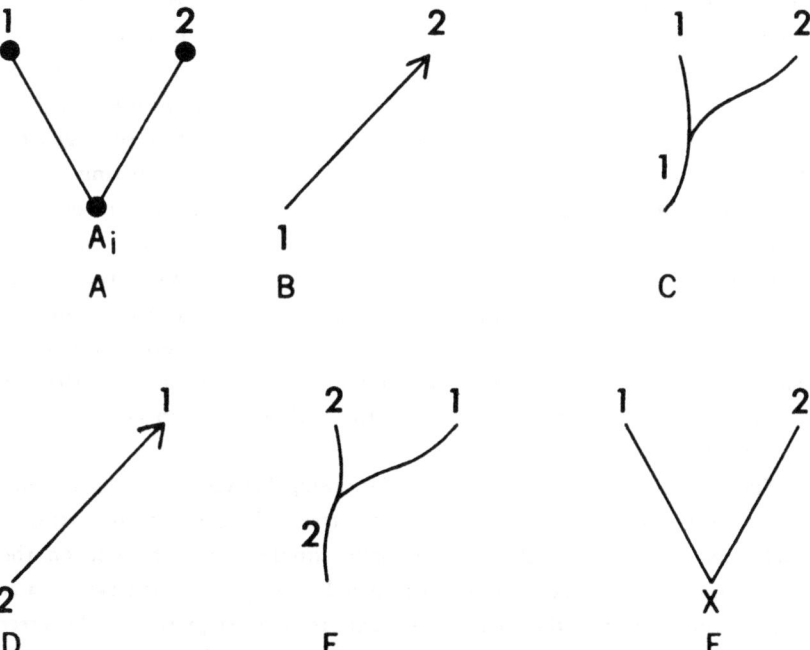

Figure 6. The generalized "ancestor tree" ("A-tree") and its five specific forms: (B and C) $i = 1$; (D and E) $i = 2$; (F) $i \neq 1$ or 2, that is, $i = X$, where X is unknown but is neither taxon 1 nor taxon 2. In B and D, the ancestor becomes extinct by transformation into its descendant. In C and E, the ancestor persists alongside its descendant.

the majority of the possible patterns of ancestry and descent as long envisioned in evolutionary theory. Figures 6B and 6D denote phyletic evolution, where one taxon is transformed into another, and the ancestor becomes "extinct by transformation." (Clearly, this situation is inconceivable for genera and taxa of higher rank.) Phyletic gradualism and saltationism are two mechanisms commonly postulated to produce this form of phylogenesis. The situation in figures 6C and 6E, in which the ancestor survives as a contemporary of its descendant, is implicit in much of speciation theory and is the expected outcome of Type $b$ allopatric speciation as well as parapatric and sympatric speciation, as recently reviewed by Bush (1975). Finally, the situation depicted in figure 6F, in which the ancestral taxon divides, thus becoming extinct by transformation into two daughter species, conforms to Bush's Type $a$ allopatric speciation. Other possibilities of course exist since speciation may be polychotomous whether or not the ancestor persists. It is even conceivable that, in some circumstances, polychotomous speciation could be the expected outcome. However, it is well to remember at this juncture that the conformity between evolutionary theory and the actual form of phylogenetic trees, although interesting and ultimately necessary, need have no direct bearing on the formulation of a strategy to reconstruct phylogenetic history (*contra* Eldredge and Tattersall, 1975). When constructing cladograms and trees, we have two or more taxa (actually, a minimum of three taxa are necessary for devising and testing alternative cladograms) which we think might have this kind of relationship. When only two taxa are involved, as in figures 6A–E, in which the basic and most general form of a phylogenetic tree is considered, polychotomies are by definition irrelevant. However, cladograms that depict polychotomies among species, and that are usually considered to be in need of further anlysis, may in fact represent polychotomous speciation.

Thus there is a direct topological relationship between a cladogram and the most general form of a phylogenetic tree. The task of converting a cladogram to a tree is, then, superficially simple: For each node on the cladogram, specify a taxon. That taxon may be either one of the two taxa at the terminal points of the clades or a third taxon either present as the sister group of taxon $(1+2)$ or as yet unknown. But how in fact do we go about converting a cladogram into a tree? It is meaningless simply to write "$A_i$" at each branching point on a cladogram; how do we actually go about specifying the value of "$i$"?

First we must realize that we have added a rather large assumption to our analysis. We have assumed that there actually is a pattern of ancestry and descent for the taxa we are studying, i.e., that some of the taxa under study are ancestors and some descendants in a direct, specific sense. All could, after all, be the twigs of the tree. Moreover, there could be any number of "intervening" taxa which we know nothing about. Trees are evolutionary event-o-grams, making specific statements about actual ancestry and descent. Campbell (1975) criticized Eldredge's (1973) cladogram on, among other things, the grounds that Eldredge failed to consider other taxa beyond the Appalachian faunal province which (undoubtedly) are more closely related to specific taxa on Eldredge's cladogram. It is at the level of analysis of phylogenetic trees that this criticism is relevant.[5] Thus it behooves anyone wishing to construct a tree to secure all known taxa within the known bounds of the geographic and stratigraphic distributions of taxa that could conceivably be a part of the pattern of ancestry and descent. Thus, although we never know if we have all the relevant taxa (actually we know we never do), some situations seem more amenable to tree construction than others; i.e., the assumption that there is a pattern of ancestry and descent among available taxa seems more reasonable in certain situations. Such a situation might pertain for genera of a distinctive family endemic to a depositional basin (e.g., the Synphoriidae within the Appalachian province; (Lespérance, 1975; Lespérance and Bourque, 1971) or, in the Recent biota, to species exhibiting parapatric or allopatric vicariant patterns of distribution.

Let us be willing to make this assumption. We then consider, for any two taxa, whether $i = 1$ or $i = 2$. To carry this out, we note than an ancestor must be considered to be primitive or equally derived in all specifiable attributes, relative to its descendant. This is true by definition but in practice requires an additional assumption, namely, that character reversal is impossible (N.

---

[5] Campbell's (1975) main criticism of "cladism" is that the work of one of its avowed proponents (Eldredge, 1973) is faulty. Campbell shows that *Phacops iowensis* may in fact be more closely related to *Phacops (Phacops) rana* than it is to *Phacops (Viaphacops) cristata*, which is contrary to Eldredge's conclusion. However, Campbell's methodology for demonstrating Eldredge's possible misinterpretation of relationships consists of asserting an alternate pattern of synapomorphy, invoking parsimony (1975, p. 91) and preparing the holotype of *P. iowensis*, exposing the hypostoma (additional data!), and asserting a resemblance to the hypostoma of *P. rana* which he considers significant (read "synapomorphic"). He thus uses the very methodology he criticizes to challenge Eldredge's conclusions but somehow then decides that Eldredge's putative error indicates that the methodology he used ("cladism") is unusable.

Platnick, personal communication, 1977; see also Engelmann and Wiley, 1977, for a different view of the use of autapomorphies to falsify trees.). Consider the alternative. If we allow character reversal—that is to say, the case in which an apparent evolutionary novelty in an ancestor reverts to the primitive state in its descendant—we are free to make ancestors out of anything, thus learning nothing and defeating our purpose. Although we would presumably stop short of deriving bivalves from brachiopods, nonetheless we are free to specify ancestors strictly on any criterion, e.g., stratigraphic position.

Thus we examine a cladogram and determine which of two taxa is the relatively more plesiomorphic (one taxon is almost invariably more plesiomorphic than the other in any sister group). If we do not allow character reversal, we can hypothesize either taxon 1 or taxon 2 to be ancestral to the other if we determine it to be plesiomorphic or at least equally (syn)apomorphic in all character states that we can compare. Discovery of a single autapomorphy in the putative ancestor immediately falsifies the hypothesis, but again, only if we are willing to assume that character reversal does not take place.[6] For example, Eldredge (1972) described the genus *Parabouleia* on one specimen which seemed to be an excellent morphologic intermediate between *Metacryphaeus* (primitive) and *Bouleia* (derived). Further work, based on over 50 additional specimens, revealed that *Parabouleia* is confined to the lowest part of the Devonian sequence in Bolivia, whereas *Bouleia* is confined to the upper part of the section. However, Eldredge and Branisa (unpublished manuscript) have described the "eye" of *Parabouleia*, finding that it is actually reduced to a slender and apparently lensless stalk, whereas *Metacryphaeus* and *Bouleia* both have normal, well-developed schizochroal eyes. Thus they rule out any possible direct ancestor–descendant relationship between *Parabouleia calmonensis* and *Bouleia dagincourti*. The hypothesis was falsified. Paleontologists for years have been making similar statements, such as, "This taxon is too specialized to be the actual ancestor." No such potential falsification is

---

[6] Many paleontologists claim that specified ancestors are merely best approximations. "I never said it was the *actual* ancestor" is the common cry. However, there is no formal difference between labeling a taxon "ancestral" or merely the "plesiomorphic sister group," except that specifying ancestral status is more specific and downright misleading to those of us who tend to take these statements seriously. I assume, in this discussion, that those who identify ancestors are not kidding; if they are, I fail to see why they persist in their activities.

possible if character reversal is allowed. However, character reversal is a definite possibility in evolution! Thus we have a serious objection, on methodological grounds, to the construction of phylogenetic trees: They are untestable statements unless we are willing to accept an assumption many of us might find overly restrictive and unrealistic.

Assume that we determine that either taxon 1 or taxon 2 is in fact plesiomorphic or equally synapomorphic in all characteristics surveyed, with respect to the other taxon. We then hypothesize it to be the ancestor. We are then faced with determining which of the two possible forms of ancestry and descent (i.e., wholesale transformation or speciation) is the least unlikely. The first step is an obvious one: Simply check the stratigraphic positions of the taxa. We would expect, naturally, that ancestors would occur in sediments older than those enclosing descendants. If we find such a congruence, we are entitled to feel that the hypothesis of ancestry and descent has been corroborated, at least to some extent. But the corroboration is not all that strong, simply because, if we find the putative ancestor always to be younger than its putative descendant, we still might not wish to throw out the hypothesis. In other words, the fossil record being what it is, it is all too possible to find the components of a phylogenetic sequence out of order. For example, Stensiö (1959) stuck to his guns in his analysis of primitive versus derived states of emargination of the shoulder girdle of euarthrodires, even though his scheme ran exactly counter to the stratigraphic sequence of these fish. Of course, if we are willing to assume that the stratigraphic sequence is always so complete that, in effect, it "never lies," we may then wish to read phylogeny directly from the rocks, and even our most basic conclusions concerning the distribution of primitive versus derived states will be dictated solely by relative stratigraphic position. In such a system, the entire comparative analysis advocated up to this point in this paper would be wholly superfluous. But what, then, would be the methodological yardstick which would allow us to draw "plausible" trees for some things (e.g., among phacopid trilobites) but not others (e.g., between clams and brachiopods)? We're trapped—some sort of prior comparative analysis, however unconscious, underlies even the most dogmatic attempts to reconstruct phylogenies by "chasing fossil oysters up a single cliff" (to borrow a phrase from Ager, 1973). The preceding discussion is merely an attempt to make that necessary, prior analysis explicit. Like it or not, no one in fact brings stratigraphic evidence to play in phylogeny reconstruction much before the point advocated in this discussion.[7]

Stratigraphic and (to a lesser extent) geographic distributions play an important role in evaluating the two possible forms of ancestor–descendant relationships. If we find a succession of two or more taxa whose stratigraphic ranges do not overlap and are in conformity with the pattern of synapomorphies and if we are satisfied that no close, "intervening" taxa are known elsewhere, we might conclude that phyletic transformation (figs. 6B and 6D) is the least unlikely hypothesis. This is the case with Stenzel's (1949) successional speciation of oysters of the *sellaeformis* stock. If, however, we find stratigraphic overlap between ancestor and descendant, we are forced to accept a splitting (speciation) event as the most likely situation. Such formulations can be refuted—in the former case by finding stratigraphic overlap and in the latter by showing that the overlapping (upper) portion of the ancestor's range actually represents some other taxon. But the best way to test these hypotheses is by refutation of the original cladogram. Specific hypotheses concerning actual modes of putative evolutionary events are most difficult to test decisively, since the relevant data are exceedingly difficult to procure.

In the final case, where neither taxon 1 nor taxon 2 appears to be a viable candidate for ancestry of the other, we then consider the fifth possibility: $i \neq 1$ or 2. Figure 7 illustrates the procedure. Figure 7A shows the cladogram where taxa 1 and 2 comprise a sister group, the sister group of which is, in turn, taxon 3. We first investigate whether or not taxon 3 might be the ancestor of taxon $(1 + 2)$, as expressed in the general case of figure 7B. But it is immediately obvious that taxon 3 cannot be the ancestor of both taxa 1 and 2, because the cladogram indicates the presence of at least one synapomorphy in taxon $(1 + 2)$ which must have been present in their immediate common ancestor and which, according to the cladogram (fig. 7A), is absent in taxon 3. In fact, only a trichotomy (fig. 7B), in which taxon 3 is primitive or equally apomorphic with respect to both taxon 1 and taxon 2

---

[7] Gingerich (e.g., 1967a) has spoken for many paleontologists when he advocates an approach to phylogeny reconstruction in which stratigraphic position plays the key role in determining what gave rise to what (i.e., stratophenetics) only when there is a "dense and continuous" fossil record with "good" stratigraphic control. Aside from the "How good is good?" objection, there seems a reasonable case to be made for having a methodological approach to phylogeny reconstruction that embraces all particular examples, not just special cases. McKenna, Engelmann, and Barghoorn (1977, p. 234) have recently made these and similar points with regard to the stratophenetic approach.

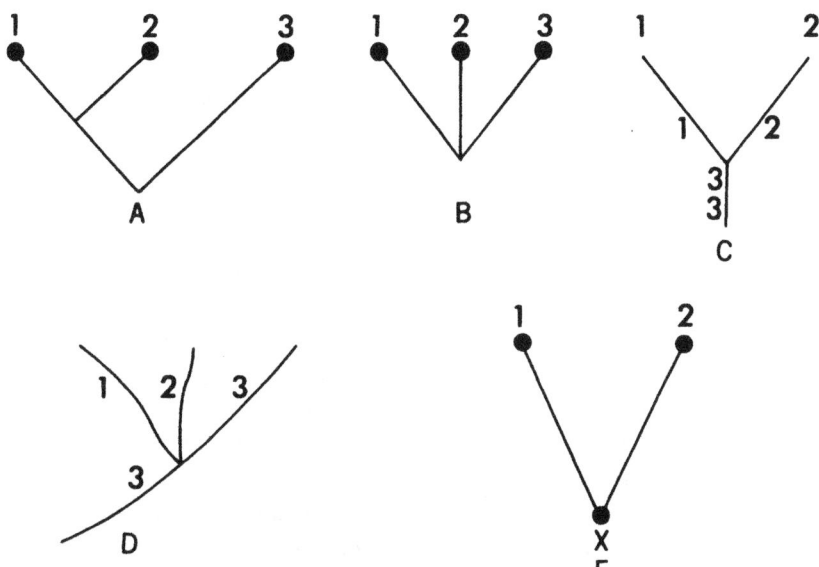

Figure 7. Use of a three-taxon statement to identify an ancestor, where $i \neq 1$ or 2. (A) Cladogram, where taxon 3 is sister group of taxon (1 + 2). (B) Trichotomous cladogram. (C and D) Two specific forms of the hypothesis $i = 3$ based on trichotomous cladogram; in C, taxon 3 divides into descendants taxon 1 and taxon 2, whereas in D, both taxa 1 and 2 are derived by speciation from taxon 3, which persists. (E) Falsification of $i = 3$, in which case we arrive back at the "X-tree," where $i = X$, and X is not taxon 1, 2, or 3.

but is distinguished from them by the presence in taxon 1 and taxon 2 of at least one autapomorphy apiece, presents a situation where a third taxon on a cladogram can be considered a possible direct ancestor of two other taxa. In such a case, we evaluate taxon 3 as was done above for taxa 1 and 2. If taxon 3 is acceptable as an ancestor, we then choose between the trees of figures 7C and 7D, depending upon distributional information. If we reject taxon 3 as the ancestor, we are left, by elimination, with the "X-tree," where "X" is an unknown ancestor. An X-tree says that we believe an evolutionary event to have taken place such that taxa 1 and 2 had an immediate common ancestor and that, although it cannot be identified as any known taxon, the ancestor is neither taxon 1 nor taxon 2 nor, for that matter, taxon 3).

Thus we see that the construction of phylogenetic trees in their most general form (i.e., simple identifications of taxa—real or hypothetical—at branching points on cladograms) requires two assumptions: (1) that an actual evolutionary event has taken place that does link the taxa at hand and (2) that no character reversal has occurred. The first assumption is required to lend any meaning whatsoever to a tree. If we are dealing only with some sort of "quasi-ancestor," we could be content with identification of plesiomorphic sister taxa and drawing the cladogram incorporating the stratigraphic ranges of the component taxa (see fig. 8A). A tree would then have nothing more in the way of information than a cladogram and would offer the serious disadvantage of beguiling the unwary into supposing that quasi-ancestors are the real McCoy. Thus we need the first assumption merely to convey truly added, new information in a tree when compared with a cladogram.

We need the second assumption simply to confer a degree of testability, i.e., to have a means of rejecting alternative trees. If we allow character reversal, then in the case of figure 6, we would never be able to choose whether taxon 1 is the ancestor of taxon 2 or the reverse. Further, and more to the point, there would be no objectively determinable limit to the morphological gaps we might leap merely to specify an *ad hoc* ancestor. Without these two major assumptions, we might just as well give up constructing trees. Finally, additional assumptions about the nature of the fossil record and evolutionary processes must be made in order to pick a specific form of tree once a putative ancestor is decided upon.

If we might as well give up making phylogenetic trees without these assumptions, we should seriously consider the possibility that the assumptions are not worth it. After all, both are hard to swallow. But I think they are worth it if paleontology is ever to be of any use in coming to grips with evolutionary processes. We study process through pattern; i.e., we predict, from our theories of process, what patterns ought to appear in the fossil record. Of course, we tend to find what we expect, but there are enough other people around with different expectations that this is not necessarily a serious difficulty. Detailed trees, i.e., hypotheses of actual ancestry and descent, underlie both Gingerich's (1976b) examples of phyletic gradualism and Eldredge and Gould's (1972) notion of punctuated equilibria, for instance. The debate that these and other contributions have engendered has at the very least served to sharpen our ideas on the relationship between paleontology and evolutionary theory and has rekindled interest in evolution on the part of many paleontologists. Cladograms alone would not have sufficed; it takes a tree—an explicit, specific hypothesis of ancestry and descent—to provide fuel to the evolutionary fire. So, there seems to be some utility to

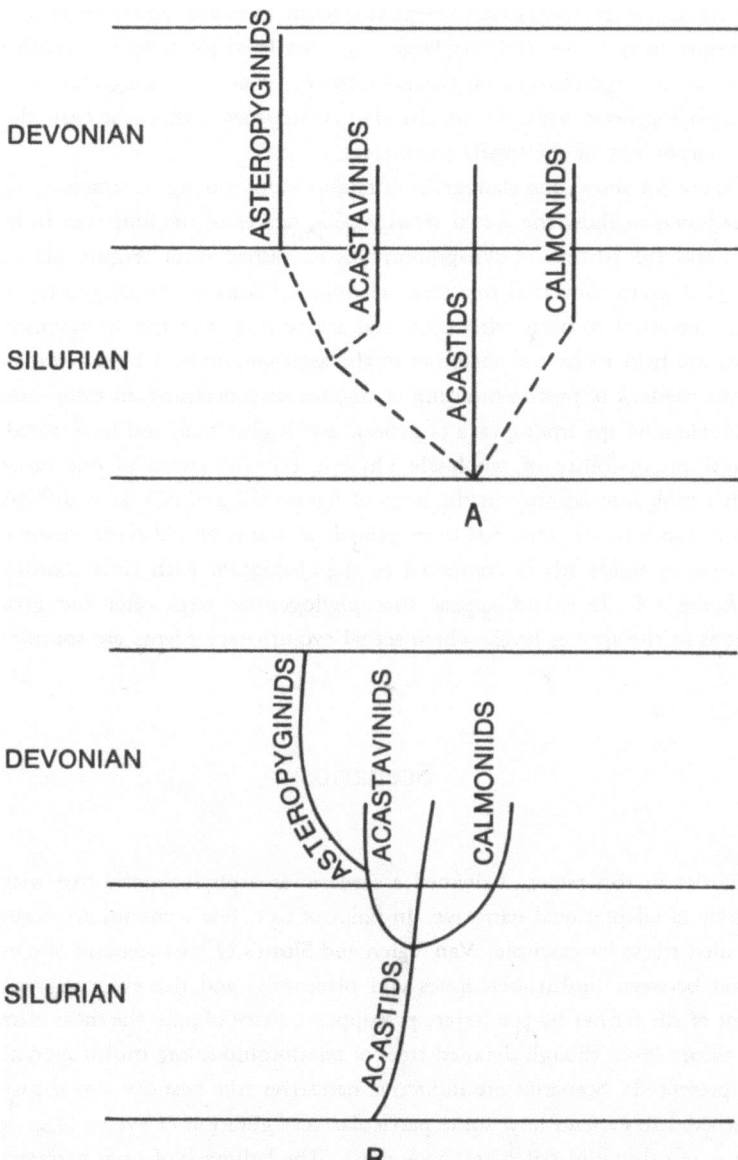

Figure 8. (A) Cladogram of relationships among Acastacea (fig. 4D), with geological ranges of the four component taxa. (B) A phylogenetic tree drived from the cladogram.

phylogenetic trees, and they remain a proper goal of phylogenetic analysis. But let us not deceive ourselves—phylogenetic trees are replete with dubious assumptions and take us a very large step away from the simple, testable patterns of the distribution of synapomorphies, that is, cladograms. Hence, with phylogenetic trees, we are already two steps away from the basic data— the morphology of the fossils themselves.

Figure 8A shows the cladogram of relationships among Acastacea of figure 4, redrawn to show the actual stratigraphic ranges of the four taxa in heavy lines and the pattern of synapomorphies in dashed lines. Figure 8B shows this cladogram converted to a tree, wherein the acastids are alleged to be directly ancestral to both calmoniids and acastavinids and the acastavinids, in turn, are held to be the ancestors of the asteropyginids. I have already discussed the lack of precise meaning of statements concerning ancestor–descendant relationships among taxa of generic and higher rank and have noted the logical impossibility of wholesale phyletic transformation of one taxon of higher rank into another in the form of figures 6B and 6D. It is difficult in this instance to see what has been gained, in terms of additional insight, in the tree in figure 8B as compared to the cladogram with time coordinates in figure 8A. It would appear that phylogenetic trees offer the greatest insight at the species level, where actual evolutionary events are specified.

# Scenarios

Earlier in this paper, I defined a scenario as a phylogenetic tree with an overlay of adaptational narrative. In point of fact, few scenarios are based on detailed trees; for example, Van Valen and Sloan's (1966) scenario of competition between multituberculates and placentals, and the eventual replacement of the former by the latter, presupposes a tree of only the most elemental nature (even though detailed trees of relationship *among* multituberculates are presented). Scenarios are inductive narratives (the best are also seductive) concocted to explain how some particular configuration of events (e.g., evolution of calmoniid trilobites) took place. The hallmark of such narratives is the analysis of the adaptive significance of evolutionary changes in size, form, and structure. Such narratives are engrossing and concern themselves with the application of what we think we know about evolution to the real world, as preserved in the fossil record. But they are mostly fairy tales constructed of a maze of untestable propositions concerning selection, function, niche utilization, and community integration, and alas, do not generally represent good science.

Yet there is a tendency to accept scenarios as *the* correct mode of formulation of statements on evolutionary history. There are two ways to improve scenarios: (1) by basing them more explicitly on trees and (2) by eliminating some of the more purely speculative elements from them. Basing scenarios more explicitly on detailed trees will help to pinpoint patterns of distribution of monophyletic groups in space and time and will yield more precise notions of ancient biogeography and the development of biotas through time. Figure 9 shows such a scenario—a phylogenetic tree of the major groups of Phacopina with geographic and gross adaptational information added. The diagram summarizes the notion that, within the entire Suborder Phacopina, there are two obvious, extreme morphological forms: "phacopiform" and "dalmanitiform." "Acastiform," harder to characterize, is intermediate between the two (as evidenced by the historical confusion over the proper placement of the acastids—Are they phacopids or dalmanitids?). No statement is made about the adaptive significance of these three morphologically defined groups. We merely assume, under the neo-Darwinian paradigm, that the structures are adaptive. Similarly, natural selection may be assumed, under the same neo-Darwinian paradigm, to be the effective process underlying the sequence of change within these three morphologies and the apparent adoption of these different morphologies by different groups. But any explicit statement about mode or rate of selection would be idle and fatuous, especially at this level, and probably for 99.99 percent of the fossil record (though see Sambol and Finks, 1977, for a stimulating study of selection in Cretaceous oysters). Adaptation and, especially, natural selection have no real value in the elaboration of most scenarios and are two examples of excess conceptual baggage which could be dropped for the improvement of scenarios.

Returning to the scenario of figure 9, by dividing the tree into these morphologic components and noting the geographic distribution of taxa, we find a pattern of vicariance. For example, the asteropyginids are interpreted to be essentially dalmanitiform Acastacea and are characteristic of the Boreal Old World region, whereas some calmoniids (e.g., *Metacryphaeus*) become the dalmanitiform Acastacea of the Austral Malvino-Kaffric faunal province.

Thus scenarios can be more informative than mere recitations of what

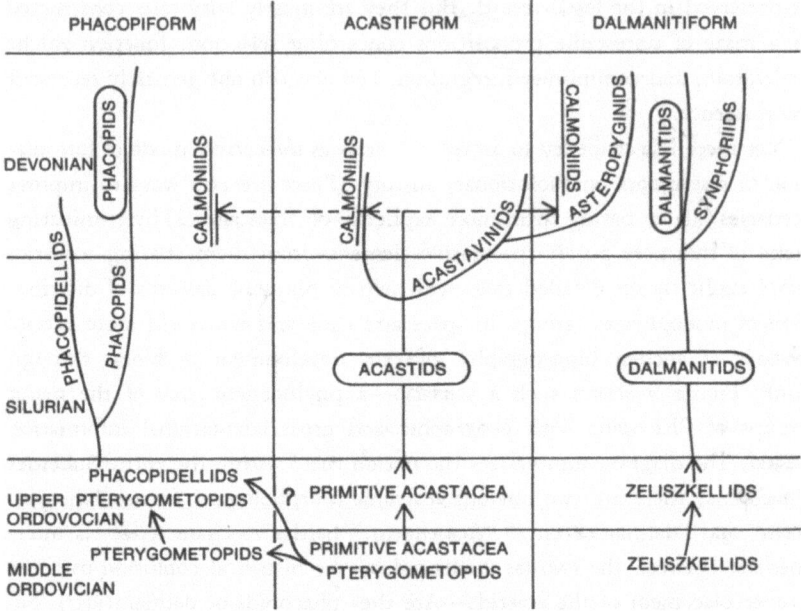

Figure 9. A scenario of the evolution of the Phacopina. Solid lines indicate phylogenetic relationships; calmoniids of different adaptive "types" are indicated by dashed lines. Adaptive "types" are indicated at the top of the diagram. Geographic distribution is as follows: Taxa in roman type are boreal, referring to the Old World and Appalachian provinces, and including, ironically, Australia; underlined taxa are austral, i.e., the Malvino-Kaffric faunal province, including South America (south of Colombia and Venezuela) and sub-Saharan Africa; cosmopolitan taxa are circled. For further details, see text.

group occurs where. They are simplified models of the basic features of the evolutionary history of a group which can serve to stimulate further investigation. As such, they are easily capable of testing and refutation. The scenario of figure 9 would be refuted if, say, a true asteropyginid were to be found in the Malvino-Kaffric faunal province (as Baldis, 1972, in fact claims). But the problem here is that we *expect* scenarios, even more than trees, to be wrong in detail. "Oh, well" we say, "just erase that part of it and make it right." This *ad hoc* changing of hypothesis content rightly infuriates strict adherents of hypothetico-deductive methodology in science and is the feature of scenarios that most horrifies them. But I would suggest that the main utility of scenarios is that they are capable of giving us ideas—lower-level hypotheses we might very well be able to test (and which we might not have formulated without the scenario)—which force us to stretch our imaginations. For instance, the scenario of figure 9 asserts that all (but the original) "invasions" into both phacopiform and dalmanitiform modes were (1) restricted to the Devonian and (2) accomplished by various Acastacea (as a real monophyletic group) and that, once they appeared, neither Phacopacea nor Dalmanitacea deviated from their basic form. It predicts, therefore, that any "pseudo-phacopid" will prove to be a Devonian acastid and never a dalmanitid.

Thus, as a heuristic device, scenarios seem to be useful. They are certainly more fun to construct than a mere tree or a dry cladogram. But by the time we reach the next step of complexity of the scenario level, we are frequently so far removed from the primary data base concerning relationships and have added so many additional assumptions (e.g., subjective characterization of "phacopiform" and "dalmanitiform" in fig. 9) that we cannot take them as serious scientific statements. But as long as we explicitly realize how we build scenarios and what their status is as scientific propositions, we should continue to build them, hoping to find them scientifically, as well as spiritually, uplifting and rewarding.

# Conclusions

I conclude from the foregoing that cladograms, phylogenetic trees, and scenarios, which are linked together in a hierarchy of complexity, are all proper goals of phylogenetic research in paleontology. I merely urge that, in their construction, we proceed from the simple to the complex and remain aware of the *ad hoc* assumptions introduced with each successive level of analysis. This is only common sense.

# Acknowledgments

Gareth Nelson first made clear to me the distinction between cladograms and trees. I acknowledge with thanks personal conversations with him and with Eugene Gaffney, Norman Platnick, and George Engelmann, each of whom also permitted my perusal of an as yet unpublished manuscript. I thank the following for their comments on the first draft of this paper: J. Cracraft, B. N. Haugh, N. D. Newell, N. Platnick, B. Schaeffer, and T. J. M. Schopf.

# References

Ager, D. V. 1973. *The Nature of the Stratigraphical Record*. 114 pp. John Wiley, New York.
Baldis, B. 1972. Trilobites devonicos de la Sierra de Santa Barbara (Provincia de Jujuy). *Ameghiniana* 9:35–44.
Brundin, L. 1972. Evolution, causal biology, and classification. *Zool. Script.* 1:107–20.
Bush, G. 1975. Modes of animal speciation. *Ann. Rev. Ecol. Syst.* 6:339–64.
Campbell, K. S. W. 1975. Cladism and phacopid trilobites. *Alcheringa* 1:87–96
Campbell, K. S. W. 1977. Trilobites of the Haragan, Bois d'Arc and Frisco Formations (Early Devonian), Arbuckle Mountains Region, Oklahoma. *Okla. Geol. Survey Bull.* 123, 227 pp.
Clark, W. E. Le Gros. 1959. *The Antecedents of Man*, 374 pp. Edinburgh University Press, Edinburgh.
Clarkson, E. N. K., Eldredge, N., and Henry, J.-L. 1977. Some Phacopina (Trilobita) from the Silurian of Scotland. *Palaeontology* 20:119–42.
Eldredge, N. 1971. Patterns of cephalic musculature in the Phacopina (Trilobita) and their phylogenetic significance. *J. Paleontol.* 45:52–67.
Eldredge, N. 1972. Morphology and relationships of *Bouleia* Kozlowski, 1923 (Trilobita, Calmoniidae). *J. Paleontol.* 46:140–51.
Eldredge, N. 1973. Systematics of Lower and Lower Middle Devonian species of the trilobite *Phacops* Emmrich in North America. *Bull. Amer. Mus. Nat. Hist.* 151:285–338.
Eldredge, N. 1979. Alternative approaches to evolutionary theory. *Carnegie Mus. Bull.*, no. 13, in press.
Eldredge, N., and Branisa, L. Calmoniid trilobites from the Lower Devonian *Scaphiocoelia* Zone (Lower Devonian) of Bolivia, with remarks on related species. Unpublished manuscript.
Eldredge, N., and Gould, S. J. 1972. Punctuated equilibria: An alternative to phyletic gradualism. In T. J. M. Schopf (ed.), *Models in Paleobiology*, pp. 82–115. Freeman, Cooper and Co., San Francisco.
Eldredge, N., and Ormiston, A. 1979. Biogeography of Silurian and Devonian trilobites of the Malvino-Kaffric Realm. In A. J. Boucot, and J. Gray (eds), *Histor-*

*ical Biogeography, Plate Tectonics, and the Changing Environment.* Oregon State University Press; in press.

Eldredge, N., and Tattersall, I. 1975. Evolutionary models, phylogenetic reconstruction, and another look at hominid phylogeny. In F. S. Szalay (ed.), *Approaches to Primate Paleobiology. Contrib. Primatol.* 5:218–42. S. Karger, Basel.

Engelmann, G. F., and Wiley, E. O. 1977. The place of ancestor–descendant relationships in phylogeny reconstruction. *Syst. Zool.* 26:1–11.

Gandl, J. 1972. Die Acastavinae und Asteropyginae (Trilobita) Keltiberiens (NE-Spanien). *Abh. Senckenb. Naturforsch. Ges.* 530:1–184.

Gingerich, P. D. 1976a. Cranial anatomy and evolution of Early Tertiary Plesiadapidae (Mammalia, Primates). *Univ. Mich. Mus. Paleontol. Pap. Paleontol.* 15:1–140.

Gingerich, P. D. 1976b. Paleontology and phylogeny: Patterns of evolution at the species level in Early Tertiary mammals. *Amer. J. Sci.* 276:1–28.

Hecht, M. K. 1976. Phylogenetic inference and methodology as applied to the vertebrate record. *Evol. Biol.* 9:335–63.

Hecht, M. K., and Edwards, J. L. 1976. The determination of parallel or monophyletic relationships: The proteid salamanders—a test case. *Amer. Nat.* 110:653–77.

Hennig, W. 1965. Phylogenetic systematics. *Ann. Rev. Entomol.* 10:97–116.

Hennig, W. 1966. *Phylogenetic Systematics,* 263 pp. University of Chicago Press, Chicago.

Kluge, A. G., and Farris, J. S. 1969. Quantitative phyletics and the evolution of anurans. *Syst. Zool.* 18:1–32.

Lespérance, P. J. 1975. Stratigraphy and paleontology of the Synphoriidae (Lower and Middle Devonian dalmanitacean trilobites). *J. Paleontol.* 49:91–137.

Lespérance, P. J., and Bourque, P.-A. 1971. The Synphoriinae—an evolutionary pattern of Lower and Middle Devonian trilobites. *J. Paleontol.* 45:182–208.

McKenna, M. C., Engelmann, G. F. and Barghoorn, S. F. 1977. Review of "Cranial Anatomy and Evolution of early Tertiary Plesiadapidae (Mammalia, Primates)" by Philip D. Gingerich. *Syst. Zool.* 26:233–38.

Mayr, E. 1974. Cladistic analysis or cladistic classification? *Zeit. Zool. Syst. Evol.–F* 13:94–128.

Nelson, G., and Platnick, N. I. *Cladistics and Vicariance.* Columbia University Press, New York, in preparation.

Platnick, N. 1977. Parallelism in phylogeny reconstruction. *Syst. Zool.* 26:93–96.

Richter, R., and Richter, E. 1952. Phacopacea von der Grenze Emsium/Eiflium (Tril.). *Senckenbergiana* 33:79–108.

Sambol, M., and Finks, R. M. 1977. Natural selection in a Cretaceous oyster. *Paleobiology* 3:1–16.

Schaeffer, B., Hecht, M. K., and Eldredge, N. 1972. Phylogeny and paleontology. *Evol. Biol.* 6:31–46.

Shergold, J. H. 1966. A revision of *Acaste downingiae* (Murchison) and related trilobites. *Palaeontology* 9:183–207.

Simpson, G. G. 1961. *Principles of Animal Taxonomy.* Columbia University Press, New York.

Simpson, G. G. 1975. Recent advances in methods of phylogenetic inference. In W. P. Luckett, and F. S. Szalay (eds.), *Phylogeny of the Primates,* pp. 3–19. Plenum Press, New York.

Stensiö, E. A. 1959. On the pectoral fin and shoulder girdle of the arthrodires. *K. Svenska Akad. Handl.* 8:5–229.

Stenzel, H. B. 1949. Successional speciation in paleontology: The case of the oysters of the *sellaeformis* stock. *Evolution* 3:34–50.

Struve, W. 1959. Family Calmoniidae Delo, 1935. In H. J. Harrington et al. (eds.), *Treatise on Invertebrate Paleontology, Part O: Arthropoda* 1, pp. 0483–89. Geological Society of America, Boulder, Colo., and University of Kansas Press, Lawrence, Kansas.

Van Valen, L., and Sloan, R. E. 1966. The extinction of the multituberculates. *Syst. Zool.* 15:261–78.

# Cladistics: Is It Really Different from Classical Taxonomy?

**A. J. Boucot**

The organizers of the symposium have asked me to provide a short commentary on the five papers. In plain words, it seems to me that the late Willi Hennig and his followers have rediscovered the wheel; they have rediscovered taxonomy more than 200 years after its formal, Linnaean birth. After stripping away the jargon of "morphos," "apos," "plesios," "syns," "character states," and whatnot, I can only conclude that "phylogenetic systematics," or "cladistics" (call it what you will), is nothing more or less than old-fashioned taxonomic classification so plastered over with jargon as to be unrecognizable to the casual reader. Gaffney has performed a useful service by reviewing the generalities, both philosophical and down-to-earth, that are necessary to put together reliable taxonomies, classifications, and phylogenies. But he provides no standard program for taxonomic–cladistic efforts, no rules of thumb for the tyro, for he merely formulates the common sense of all those from Aristotle to the present who have been concerned with the nature of classifications and the logic implicit in their construction. The general problem of deciding whether or not a whale or an ichthyosaur might be a fish is always with us as we try to understand more fully the taxonomy of all groups at all taxonomic levels. Arguments over the correct taxonomic position of taxa at all levels and in all groups of plants and animals, living and fossil, are merely a reflection of this old, old problem, which will never totally fade away into that sunset of perfect truth and ultimate understanding.

To my mind, the principle contribution of the cladists lies in their efforts to systematize taxonomic procedures and encourage more meaningful, better-written diagnoses. Cladograms in themselves are really nothing more nor less than graphical taxonomy. Cladists demand that primitive and advanced features, as well as the different intermediate grades, be firmly recognized and systematically worked into the taxonomy—pardon me, the phylogenetic systematics. How does this differ from the efforts of the classical taxonomists? Not at all, as far as I can see. The cladists insist, as pointed out here by Eldredge, that cladograms are not family trees; let me remind you that taxonomic classifications are not family trees either. I have heard more than one highly respected, thoroughly competent taxonomist complain vigorously that it is premature to begin the construction of family trees, because of the huge gaps in our knowledge of the fossil record. Does this suggest that the taxonomist makes no distinction between taxonomic classifications and family trees? How does this differ from Eldredge's statement that cladograms precede the construction of family trees? Have taxonomists seriously suggested that family trees be constructed *before* the taxonomy of the group was worked out? Eldredge points out that taxonomic classifications *cum* cladistic arrangements are *not* family trees. I doubt that many taxonomists have maintained now or in the past that taxonomic classifications were synonyms of family trees.

Eldredge presents an example of cladistics in practice, which employs phacopid trilobites. One is struck with the identity between cladistic logic and that used by most of us as pure and simple, old-fashioned taxonomic discussion. Is there really any difference except for the jargon? I think not.

Taxonomy flourished as a purely nuts-and-bolts approach to rational organization long before biologists and paleontologists gave much serious thought to evolution and family trees. The holotype concept is not a new one! In essence, taxonomy is nothing more than the attempt to classify organisms, or objects, logically, in a rational manner indicating their degrees of similarity and difference. But is it really necessary (following Hennig, as quoted by Eldredge) to refer to the "heterobathmy of synapomorphy" when suggesting that hair is a feature common to all mammals, and therefore primitive? There is a time and a place for well-considered technical nomenclature as an aid to communication and understanding. The cladists seem, unfortunately, to have swallowed a rhyming dictionary rich in classic roots of all sorts; the resulting deposit has now fertilized a plague of toadstools, sprouting on our beautiful taxonomic lawn.

In any taxonomic scheme, certain features possessed by all members of a taxon are used as key features, which automatically excludes forms lacking these features. Cladistics, in common with taxonomy, faces the same basic questions again and again in trying to decide what are primitive features, advanced features, and parallel or convergent features. The cladists, in common with the taxonomists, have discovered no philosophers' stone enabling discrimination of such features. Cladists, in practice, are still reduced to becoming familiar with a particular group of organisms, working out their potential relations as objectively as possible using all the available evidence, and then producing a cladistic analysis which differs not at all from an ordinary taxonomic classification. As far as I can see, the only notable difference between the cladist and the ordinary taxonomist going about his or her business with those dusty drawers of specimens is that the cladist makes a fuss about the cerebral processes involved, presents a graphical taxonomic outline—a cladogram—and insists on interjecting references to Western philosophers of all stripes. (Thank God they have not yet discovered the Eastern philosophers, and let us hope that nobody informs them about the famous mystics of the past and their accomplishments.) But when all is said and done, I cannot see that cladistics is anything but that poor old, grubby Cinderella, taxonomy, dressed up in a snappy new outfit and riding in a cladogram drawn by I am not sure what type of organism.

Some of the cladists make a big deal out of the all too obvious fact that stratigraphic position is not a sure formula for determining ancestor–descendant relationships. There may have been a few misguided taxonomists who insisted on this position, but I fail to find it specified as any taxonomist's "law." Again, taxonomists have never insisted that their various categories, from the species on up, have any specific age content or requirement. We classify things regardless of their geologic age. If Pete Palmer [Dr. A. R. Palmer, SUNY at Stony Brook] were presented with a wriggling, squirming, dripping *Paradoxides*, I am certain that he would insist on calling it *Paradoxides*, after first having knocked it over the glabella to put it out of its misery. *Neopilina* and *Latimeria* posed no problems in taxonomic assignment because of their youth.

I suspect that few taxonomists seriously consider family trees to be the actual path followed by evolution at the species, genus, or even family, level. Most of us take a very detached view of family trees. But—and it's a big but—most of us do view family trees as an objective advance over a mere taxonomic classification. Why? Because by plotting a family tree, which is in essence our concept of systematic changes in morphology against time (actually against stratigraphic correlation), we are able to introduce geologic time into the mix. Why is this important? And why is it superior to a pure taxonomic classification? Because of the gamblers' odds, worked out over the past two centuries, that similar shapes within the same major taxonomic grouping tend to occur closer to each other in geologic time than is true for different shapes. By arraying similar morphologies against stratigraphic correlations, we have a better chance of at least approximating that elusive "natural" taxonomy rather than a purely nuts-and-bolts taxonomy. Again, few of us would maintain that such family trees, at any level, represent the actual path taken by evolution, but we do maintain that they are more likely to represent an approximation of that path if we have accumulated enough data and if—following Bretsky and Gingerich—our samples are adequate. Should our samples be inadequate, as is the unfortunate case for all too many groups, we are realistic enough to understand that our taxonomies and our family trees are at best diagrammatic and at worst one of an extremely large number of alternate possibilities from which we could have chosen.

Both Eldredge and Cracraft strongly suggest, without really documenting their suspicions, that the bulk of the old-fashioned practitioners of the taxonomic art unjustifiably color their reconstruction of family trees in terms of stratigraphic position. Until they can really prove their point by citing numbers of papers for and against their position, I can only view their fears as phantoms conjured up in support of the cladistic jargon. Cracraft goes to some lengths in his assault on whole phalanxes of taxonomic straw persons.

Eldredge and Cracraft have rightly emphasized that relative stratigraphic position is an invalid taxonomic criterion, cladistic criterion, or basis for constructing a family tree. However, one would do well to heed the seat-of-the-pants "feel" most paleontologists have gained over the years regarding the fact that close taxonomic relationships at the species, genus, and sometimes even the family level tend to involve relatively small time ranges. This is admittedly a statistical type of thing. Few taxonomists are very keen to link up an Ordovician species or genus with one in the middle Miocene unless the evidence of morphology, community affinity, biogeography, and goodness knows what else, are monumentally overwhelming. In other words, with big time gaps between taxa of some potential relation, we can gamble that the large gap lessens the possibility of close taxonomic affinity, particularly at the species level, but does not totally remove it. Conversely, as we close up the time gap, time has increasing utility as a guide to possible affinity. All of this is aimed at those all-too-common, normal situations in which the sampling program advocated by Gingerich is not realized. Normally, we deal with a group of miscellaneous, chiefly small, samples, scattered in both time and space (the trilobite-based example used by Eldredge in this symposium is all too typical in this respect—scattered specimens, weak stratigraphic documentation, typical museum collections). So we do the best we can by employing gross morphology, correlation, etc., while realizing that many of our taxonomic conclusions may be found to be wrong.

It is very heartening to read that Eldredge, who appears to be a devoted cladist, heartily supports the construction of family trees. It would be, in my opinion, and as he rightly points out, dereliction of our duty to the evolutionary concept if we refused to risk a family-tree construction because it might be somewhat in error. Should we withdraw from considering the various family-tree possibilities, we would become mere describers and classifiers of objects *ad infinitum*. However, Eldredge's comment that family trees offer the "greatest insight" at the species level is aimed at a peculiar definition of "greatest insight." Surely one of mankind's greatest achievements, following in Darwin's footsteps, has been to suggest that birds, via *Archaeopteryx*, are descended from reptiles, with both better and less well documented examples taken from the various kingdoms. Gingerich, fortunately, labors this point still more.

Eldredge's statements about "scenarios," although praiseworthy in intent, are still a bit confused. It might have been better for Eldredge to have reviewed some of the far better known examples from the marine environment where the complete story of changing taxonomy, changing and fixed features of community history, together with the changing and permanent features of the biogeography have been interpreted in one way or another.

It is not surprising to this taxonomist to learn from Eldredge that taxonomic relationships can be suggested by the study of ontogenies. The latter half of the twentieth century is apparently catching up with the latter half of the nineteenth century. Here again, it is difficult to see that cladistics, far from being new-fangled and original, is anything more than old-fashioned taxonomy with a shiny new coat of jargon. Bretsky, too, quite rightly supports the use of ontogenies in working out taxonomies, and eventually family-tree relationships.

Eldredge makes the point that it is difficult to weight characters as to which are most important in terms of primitiveness—an old, old taxonomic dilemma. This is clearly a situation in which only the experienced taxonomist who has studied a large sample is in a position to evaluate the various possibilities. There is no purely logical way out of this question. Weighting the two classes need not be the proper procedure either. What is needed here is the old-fashioned, but thoroughly experienced, taxonomist. It will be interesting to see how the inexperienced armchair theoreticians arrive at ultimate truth. One can only wish them luck, they will need it.

In many ways, the cladists, in common with the old-fashioned taxonomists, show little concern for anything other than pure morphology. Why else would Eldredge call for ontogeny? Harper (1976) has emphasized, among other things, the importance of studying the biogeographic distribution of similar, or potentially related, morphologies through time as well as carefully reviewing the taxonomic *cum* cladistic ground rules. In addition, it is crucial to study the community affinities of given morphologies through time (Boucot, 1978). It turns out that communities are remarkably stable entities through time. Both the abundant and the cosmopolitan species, in most instances involving benthic marine invertebrates, persist through significant time intervals commonly measured in fives or even tens of millions of years. The rarer, commonly endemic species may be interpreted to have evolved phyletically while persisting in the same evolving community group. These evolutionary changes, whether interpreted as phyletic, which I prefer, or as allopatric, as preferred by Eldredge and Gould (1972; Boucot, 1978) suggest, when viewed in the community context, that taxonomic affinities, i.e., cladistic relationships, may be commonly considered as noncladogenetic. Real branching of one species from another does not occur since the two species are not found sympatrically except in gradational situations—which are rare as judged by carefully studied, adequate samples. This type of evolution may be viewed as phyletic or zig-zag (Henningsmoen, 1964) but not as cladogenetic in the commonly accepted sense. Eldredge appears in practice to decry considerations other than the purely morphologic, as would many taxonomists, but if one is trying to eliminate the ambiguities implicit in parallel evolution (discussed by Harper, 1976, and also of concern to Eldredge), then it is critical that all evidence, morphological or not, be employed in constructing our taxonomies *cum* cladogrammic graphical taxonomies. A plea here for including biogeographic and community information as a routine part of the species and genus description may not be out of place!

In my opinion, the most important contribution made by this symposium is the insistence, well supported by actual examples, that the working out of reliable taxonomy (ultimately, of the family trees, which are never any better than the taxonomy) is highly dependent on the quality of the sampling. For this emphasis, we can thank Bretsky and Gingerich. A taxonomy *cum* cladistic analysis is only as good as the sampling on which it is based. Closely spaced, carefully studied samples from well-measured sections, subjected as well to careful community and biogeographic analysis, give rise to reliable taxonomies, and these in turn provide a sound basis for family-tree speculations and maybe even Eldredge's scenarios.

Thus far, we have chiefly discussed the specific level in taxonomy, that is, cladistics. At the supraspecific level, as with so many evolutionary questions involving fossils, one finds that George was here earlier. The terms quantum evolution and phyletic evolution (Simpson, 1944, 1953) cover the ground very nicely. The cladists do make a useful point in reemphasizing that family trees for supraspecific levels, and particularly for suprageneric levels, are plagued by missing links at the specific level. So do the creationists! Most cladists studiously avoid Simpson's (1944, 1953) concept of quantum evolution, which helpfully evaluates this negative evidence at the specific level. However, it takes a pretty biased cladist to insist that most taxonomists have claimed to recognized complete species- or genus-level continuity between most families and higher categories. Some of us may have been careless in using solid lines to connect families deduced on one basis or another to be ancestral to other families, but I hope that will not be held against all of us. Many taxonomists have been very scrupulous in employing dotted lines, dashed lines, queries, and similar devices, and few of the solid-line scribers have ever suggested, if one reads the accompanying text, that they had species- or genus-level continuity between their families and higher-level categories.

Despite the obvious fact that taxonomic classification (i.e., cladistic analysis) does not depend on mode of evolution, it was surprising to me to see how much time the five contributors spent discussing modes of evolution.

206                           A. J. Boucot

However, several (notably Eldredge) do make clear that mode of evolution is not involved in setting up a taxonomy *cum* cladistic scheme. As taxonomists, we can disagree violently about evolutionary mechanisms even while agreeing on the basic, taxonomic classification of the groups in question.

The bulk of Cracraft's paper seems to fall outside the purview of the symposium; I thought we were concerned with phylogenetic models rather than modes of speciation. Possibly, Cracraft's position merely reflects the frustrations one must encounter in trying to piece together the available, all too rare, scraps of ratite bone into a coherent taxonomic and evolutionary picture; maybe the buzzards did not leave much for Joel to work with. Fortunately, few paleontologists have done more than describe, identify, name, and classify their materials. In other words, they have been practicing taxonomists. Had they lived up to Cracraft's hopes, the published record would have contained so much speculation and philosophical featherbedding that we would have had room for describing only a small fraction of the species handled to date. Finally, Cracraft puzzles me when he states that Simpson "remained essentially a Darwinian gradualist." How can one so characterize the deviser of quantum evolution? Simpson certainly is a convinced gradualist concerning species-level phyletic evolution as the common mode, but this position is based on his wide experience as a practicing taxonomist faced with day-to-day problems of assigning specific and generic terms in situations where "taxonomic indecision" is all too prevalent. One of Simpson's chief distinctions as a scientist is that he clearly articulated the experience of many and derived the appropriate generalizations from it. It seems needless to appeal to metaphysics and abstractions in trying to explain why practicing taxonomists have found gradualism acceptable. A few years' drudgery trying to identify specifically a group of closely related fossils usually convinces most practitioners of the art that although ecophenotypes may explain some of the trouble, evolutionary gradualism is also needed.

Cracraft makes the point that the way in which we construct our taxonomies, as well as the evolutionary concepts involved in the family-tree concepts derived from them, is somehow a function of our cultural heritage. Were this the case, one would expect taxonomists educated in different societies, under different political and economic systems, to approach taxonomy and family-tree construction in a different manner, which would yield very different results. I can only speak for myself, but as an older taxonomist who has imposed himself on colleagues from most of the so-called civilized countries of this world, I have found that the approaches and results of these taxonomists turn out to be remarkably similar to my own. Could it be that the nature of our materials, both living and fossil, exercises a real power over our thinking? Could it be that what you and I agree are two apparently like or unlike seashells actually appear equally like or unlike to an outsider with a different cultural heritage? Could it be that attempts to classify shapes are similar no matter what type of pledge of allegiance one recited, or failed to recite, no matter what uniform or mufti one wore as a child and young adult? I have a hunch that fossils dictate far more to us, as taxonomists and evolutionists, than some of us are willing to admit. Maybe more time with the fossils in the field and laboratory, more careful section measuring and collecting, and less time in the library with the philosophers might help.

Bretsky raises the important point of variation. She is rightly concerned with how little intraspecific variation has been documented, except for a few studies like those of Gingerich, either at one locality or over a contemporary environmental gradient. She makes the important point that, until more study of intraspecific variation is done across contemporaneous environmental gradients, as well as through time, we will have problems assessing just how evolution may have proceeded. Critical to determining modes of species-level evolution is the attempt to disentangle ecophenotypic variation paralleling contemporaneous environmental gradients from truly evolutionary, consistent, clinal changes in morphology. The presence of so much "taxonomic indecision" in the day-to-day identification of the more abundant species provides a potential measure of the presence of clinal change. However, it appears that the best one can do at present is to back away, taxonomically speaking. By "backing away," I mean studying closely related genera and subgenera from the same community group, in the same biogeographic unit through time. If the same community group has representatives of closely related, changing genera and subgenera through time, there is every reason to suspect that at least a reasonable proportion of the clinal changes we see will be of the evolutionary rather than the ecophenotypic type. In addition, of course, we could expect ecophenotypic changes through time to reverse themselves more commonly than would be the case with the evolutionary types.

Few reasonable practicing taxonomists will find cause to disagree with Gingerich's "stratophenetic" approach (although some may gag on the term).

It is obvious that a really dense stratigraphic record, restricted to the same environment as evidenced by both community-group and biogeographic analysis, is the only reliable and "final" means for determining modes of speciation. Few paleontologists during the past two centuries have, however, gone to the labor exemplified by Gingerich's studies. We should try to emulate him. How else can we accumulate the information needed to satisfy the quibblers and skeptics? Appeals to "taxonomic uncertainty" as evidence are unconvincing to those armchair theoreticians whose day-to-day experience of dull, routine taxonomy is limited. Really richly sampled sections are the answer to our problems. The evidence provided by the rare fossils will, by definition, always remain ambiguous no matter how diligent the investigator.

One important generalization may be made about family trees. This generalization is one which none of the papers in this symposium has addressed. I'm concerned here with the characteristics of communities through geologic time. As mentioned earlier, recurring associations of seashells, bones, leaves, or pollen grains, i.e., communities for the paleontologist, do not behave as random aggregates of miscellaneous taxa characterized by relative and absolute abundances which vary in a chaotic manner. There is a high degree of regularity and repeatability in the taxa present and the relative and absolute abundances of these taxa (or at least in the upper and lower limits of these abundances). Within these communities, it is also clear (Boucot, 1975) that most of the changes and substitutions of one taxon for another through geologic time consist of one species of a genus being replaced by another species of the same genus. In some instances, this substitution of one species for another goes so far that it reaches the subgeneric level and even the generic level, but almost always occurs in the same subfamily. There is no change in either the number of species present within an evolving community through time (a time sequence of parallel communities) or the total number of communities present through time in any one biogeographic unit. In other words, within any one biogeographic unit, the total number of species present tends to remain about constant. There is replacement of one species by another. The instances where replacement occurs chiefly affect the less abundant, commonly endemic forms rather than the more abundant, cosmopolitan forms (Boucot, 1975, 1978). It is also evident that the bulk of the number of species generated through geologic time has been generated within these evolving communities in the manner outlined here. The critical species involved in quantum evolution, the development of new and complex adaptations, and the formation of new community groups—in other words, those species involved in the important evolutionary events—are few in number as contrasted with the monotonous, dull changes occurring within single, evolving community groups. These within-community-group changes involve most of the species of concern to the taxonomist. The quantum-evolution, new-complex-adaptation, new-community-group species are those which should properly concern the cladist. The important point for this symposium is that cladistic analysis unaccompanied by a careful community-group analysis of the material provides a weak foundation for evolutionary conclusions. Taxonomy need not be accompanied by community ecologic understanding, but an understanding of evolution, namely, the construction of family trees, may be materially advanced by paying attention to the community affiliations through time of the species involved. In my own case, I prefer to view the within-community replacements through time of one species of a genus by another as phyletic. Even in the absence of adequate, continuous samples ("taxonomic uncertainty") and despite the common presence of information gaps for many of the rarer species doing most of the evolving, I find it hard to conclude that species about to evolve send a small delegation into an ecologic dark corner where the process takes place, followed by an instantaneous, convenient, total disappearance of the parent species coinciding with a widespread dispersal of the single, daughter taxon, plus a reversion back to the ecologic requirements of the parent taxon.

As a practicing taxonomist, one who is repeatedly put down hard for his many transgressions of both omission and commission by the patient editors of the *Journal of Paleontology*, I feel that much can be gained in testing taxonomic *cum* cladistic schemes, and the family trees deduced from them, by obtaining additional specimens from new beds and new regions and then trying to cope with the resultant problems as our little house of cards collapses. Rather this, it seems to me, than arid scholasticism and appeals to this philosopher or that, lest we become as much a scientific laughingstock as those medieval theologians who concerned themselves with how many angels could dance on the head of a pin. All of us practitioners of the world's oldest profession (Hedgpeth, 1961) have seen our favorite stories go up in smoke time after time. We try to console ourselves that each scheme is a better approximation of the ultimate evolutionary "truth" than the last house of cards. It is clear that taxonomies do keep changing, but they don't change aimlessly. Every practitioner of this ancient craft knows full well that, in general, we do seem to be getting a handle on the more abundant, more widely distributed groups. Possibly it's just a mass movement of taxonomist ants involved in moving a variety of objects from one place on the dunghill to another, or possibly it's an irresistible, vectorial movement directed toward evolutionary understanding of past life.

# References

Boucot, A. J. 1975. *Evolution and Extinction Rate Controls*, 427 pp. Elsevier, Amsterdam, Oxford, New York.

Boucot, A. J. 1978. Community evolution and rates of cladogenesis. *Evol. Biol.* 11:545–55.

Eldredge, N., and Gould, S. J. 1972. Punctuated equilibria: An alternative to phyletic gradualism. In T. J. M. Schopf (ed.), *Models in Paleobiology*, pp. 82–115. W. H. Freeman, San Francisco.

Harper, C. W., Jr. 1976. Phylogenetic inference in paleontology. *J. Paleontol.* 50:180–193.

Hedgpeth, J. W. 1961. Taxonomy: Man's oldest profession. 11th Annual University of the Pacific Faculty Lecture, p. 1–19.

Henningsmoen, G. 1964. Zig-Zag evolution. *Norsk. Geol. Tidskr.* 44:341–52.

Simpson, G. G. 1944. *Tempo and Mode in Evolution*. Columbia University Press, New York.

Simpson G. G. 1953. *The Major Features of Evolution*. Columbia University Press, New York.

# Ancestors, Species, and Cladograms—Remarks on the Symposium

## E. O. Wiley

The theme of this symposium has been how to reconstruct phylogenies. Running hand in hand with this theme, however, have been two other controversies: the mode of origin of species and the recognition versus nonrecognition of ancestors. Before examining these topics, we might first ask, What kinds of phylogenetic hypotheses shall we reconstruct? Can we recognize ancestors? What kinds of ancestors may we recognize? Can taxa of any level be ancestral or must ancestors be species? Given species, what kind of species are we talking about? And how do our species concepts influence our outlook on phylogeny reconstruction and the mode and tempo of evolution?

I will argue six major points. (1) Supraspecific ancestors, i.e., paraphyletic taxa, are taxonomic artifacts and unnatural biological entites which have no place in any system which purports to reconstruct phylogeny and draw from that phylogeny past evolutionary events. (2) Recognition of supraspecific ancestors impedes progress toward studying the mode and tempo of evolution. Therefore, (3) only species are valid ancestral taxa. (4) The evolutionary species concept (cf. Simpson, 1961; Wiley, 1978) is the species concept best suited both for phylogeny reconstruction and for studying the mode and tempo of the evolutionary process. (5) When the evolutionary species concept is applied to phylogeny reconstruction, the number of possible phylogenetic trees for any given number of taxa is equal to the number of possible cladograms for those taxa *unless* one of the following occurs: The investigator overestimates the true number of evolutionary species, there is is a true trichotomy which involves an ancestor, or the investigator works with morphotypes whose specific status is not hypothesized. (6) The differences between cladograms and phylogenetic trees are trivial, and most cladograms are "evolutionary event-o-grams" (from Eldredge, in this volume).

## Supraspecific Ancestors

Many authors recognize such statements as "The Reptilia are ancestral to both Aves and Mammalia" as scientifically meaningful. Why? In some cases, this type of statement is considered meaningful because of the concept of monophyly employed. For example, if monophyly is defined *sensu* Simpson (1961) or Ashlock (1971), then the Reptilia would be a "natural taxon" and "monophyletic." It is then understandable why such supraspecific "ancestor–descendant" hypotheses were considered relevant, critical statements about phylogeny. In other cases, such statements would seem to be meaningful if the investigator supposed that species and supraspecific taxa are not different in principle. This would seem to be the position that Bretsky (in this volume) takes when she states that inferring the "limits" of a supraspecific taxon is no different in principle from inferring the limits of a species. In other words, viewing the practice of delimitation of species and higher taxa as similar might be expanded to viewing all taxa (including species) as similar. It would then be internally consistent to conclude that since species are linked by ancestor–descendant relationships, whole genera can also be linked in the same manner.

I do not wish to argue the *internal* consistency of either argument any more than I would wish to argue against the internal consistency of the Ptolemaic system (which was, after all, more parsimonious than the Copernican system in that it contained fewer epicycles). Neither argument can be refuted because it is internally inconsistent, but either might be refuted by showing that its basic assumptions are incorrect. (The analogy to the Ptolemaic system fits here—it was refuted because the sun rather than the earth was the center of the solar system). With reference to ancestral taxa, these basic assumptions are that such "monophyletic" groups as the Reptilia are natural groups existing in nature and that species are not different in principle from supraspecific groups.

As many phylogeneticists[1] have argued, only monophyletic groups *sensu* Hennig (1966; or *sensu* Bonde, 1975) are natural groups with a real existence in nature. Paraphyletic groups (*sensu* Farris, 1974; also see Platnick, 1977a) are *a priori* supraspecific "ancestors" of their monophyletic "sister groups" (see Engelmann and Wiley, 1977). Yet if paraphyletic groups have no "real existence in nature" (Bonde, 1975, p. 293), how can they be meaningful biological entities? They cannot, and furthermore, such paraphyletic groups cannot be justified or corroborated by their adherents. This is because all characters used to "justify" or "corroborate" such paraphyletic groups are plesiomorphies, which are irrelevant for indicating group membership (Bonde, 1975). Instead, these characters are relevant as synapomorphies of one or more taxa at a higher level of universality than the paraphyletic group in question (Wiley, 1978; also see Colless, 1972). Thus paraphyletic groups have all of the characteristics of meaningless or almost meaningless sets (Ball, 1975, p. 426) and cannot be justified by any criterion other than the opinion of the particular investigator (Engelmann and Wiley, 1977).

That species are different, *in principle,* from genera, families, and other supraspecific taxa is basic to the evolutionary process as perceived today by modern evolutionary biologists. Species, whether Recent or fossil, are composed of individual organisms that form a continuum based, ultimately, on reproductive ties or, in the case of allopatric demes or a series of asexual clones, lack of evolution (Wiley, 1978). Natural supraspecific taxa are groups of separate continua that are hypothesized to have been a single continuum in the past (Hennig, 1966; also see Bonde, 1975, Wiley, 1978). Put simply, natural supraspecific taxa accurately reflect hypothesized past history (Wiley, 1976), and the supraspecific taxa which do this are monophyletic *sensu* Hennig (1966).

I would argue that supraspecific taxa are not actively evolving entities and thus cannot "give rise" to anything. Put briefly, once a species speciates, it is no longer a single evolving lineage but a series of separate and independently evolving groups. And as Lewontin (1974, p. 162) has pointed out, these groups have no special relation to each other except the relationship derived

---

[1] Ball (1975), Bonde (1975), Brundin (1966, 1968), Cracraft (1974), Engelmann and Wiley (1977), Hennig (1966, 1975), Nelson (1971), Wiley (1977).

from common ancestry. This independence is manifested on the morphological level by what Gould and Eldredge (1977) call "Wright's rule"—that the direction of a macroevolutionary trend is essentially random with respect to speciation. This would not be possible without independence of lineages.[2] I conclude that species evolve but that supraspecific taxa do not evolve, only their included species, and that species evolve independently of one another. Further, since it is reasonable to assert that taxa which do not evolve cannot give rise to other taxa, I submit that supraspecific taxa cannot be ancestors. Eldredge (in this volume) has come to the same conclusion. I conclude that hypotheses of ancestor–descendant relationships between supraspecific taxa are biologically faulty because they run counter to evolutionary theory. Thus the set logic and phylogenetic justifications for rejecting supraspecific ancestors seem to be corroborated by the biological justifications for such rejection.

I conclude that species and supraspecific taxa are different in principle. Because supraspecific groups are historical reconstructions of past evolutionary events (see Bonde, 1975; Wiley, 1976, 1978), their justification must be made in terms of their monophyly. This can only be done with the use of those characters which test the proposition that the members of the taxon share a single common ancestral species not shared by another taxon; i.e., supraspecific groups can only be corroborated with synapomorphies (see Gaffney's paper in this volume for a review of the philosophy and methodology of this approach). Species, on the other hand, are *a priori* monophyletic by their very nature. This is not to say that all species are valid; after all, one can make mistakes. An investigator does not have to justify a species as monophyletic via synapomorphies as long as it is accepted that (a) species are lineages (see Wiley, 1978) with a real existence in nature (cf. Bonde, 1975; Gould and Eldredge, 1977) and (b) species are individuals (cf. Ghiselin, 1966, 1974; also see Hull, 1976; and Wiley, 1976, pp. 9–10).

---

[2] This is not to say that the overall effect is random; it may be directional in the sense discussed by Raup, Gould, and Schopf (1973). This brings up the interesting question of identifying long-term trends as selective or random events. If the results are the same, how are we to discriminate between luck and orthoselection?

## Species Concepts

Such considerations lead us to the conclusion that supraspecific taxa are different from species in principle and that ancestors must be species. But what kind of species?

With regard to the question of phyletic gradualism versus punctuated equilibrium, it is not surprising that Gingerich (in this volume) views the gradualistic model as the major mode of speciation. His view of process is directly related to his species concept. Gingerich (in his volume, p. 48) defines a species as "an arbitrarily divided segment of an evolving lineage that differs morphologically from other species in the same or different lineages." It is curious that Gingerich then criticizes Eldredge and Gould (1972) for producing an artificially inflated fossil record which appears punctuated because they "divide the range of variation of a variable character." Gingerich terms this "typology," yet it is what Gingerich does himself in arbitrarily dividing the range of variable characters by subdividing lineages. To quote from his article, "Hence the sample . . . labeled *Pelycodus trigonodus* . . . is recognizably different, even to the unaided eye in the field, from most samples of *P. mckennai* or *P. abditus,* even though intermediate samples connect them into one continuous temporal gradient" (this volume, p. 48). I would conclude that Gingerich is a "typologist" by his own definition of that word[3] and that his criticism of Eldredge and Gould (1972) is, to say the least, misplaced. Be that as it may, we might ask a series of more critical questions. Can an arbitrary species concept be used to justify a particular evolutionary process? Or does this lead to arbitrary evolutionary modes of speciation by the very fact that arbitrary species are used? Will such arbitrary species concepts significantly alter the phylogenetic reconstruction used to deduce the evolutionary process at work? I would suggest that the answer to these questions is yes. If this is so, then serious doubts must be expressed about Gingerich's species concept and his speciation model. (For other criti-

---

[3] The word "typology" is one of the most misused words in systematics. I doubt that either Gingerich or Eldredge and Gould are typologists in the classical sense of the word. I would suggest that throwing around such labels does little to make the real issues clear or to solve controversies.

cisms of Gingerich's views, see Gould and Eldredge, 1977, and McKenna, Engelmann, and Barghoorn, 1977.)

I would submit that if we are to speculate on evolutionary processes with the idea of trying to explain the origin of species, we must have, minimally, two things, First, we need corroborated species-level phylogenies. Second, we need a species concept that reflects, as well as possible, the true number of species in our phylogeny. In other words, we must strive for the most biologically meaningful and least arbitrary species concept we can devise. Recently, I have suggested that such a concept—the evolutionary species concept of Simpson—is already available (1961; see discussion and modification by Wiley, 1978). Under this concept, a species is essentially a lineage of ancestor–descendant populations with its own identity and evolutionary fate. Species originate when they split from their ancestral lineage and persist until they become extinct via extinction of the lineage or splitting of the species into two or more daughter species which have evolutionary tendencies that are different from the ancestral species (see Wiley, 1978 for examples of nonextinction at splitting). One logical corollary of this concept is that lineages cannot be subdivided into a series of so-called ancestor–descendant "species"; one cannot define species as lineages and then call a particular species only a part of a lineage. Yet Simpson (1961, p. 165) himself helped to bury the concept of evolutionary species (a few pages after he proposed it) because he felt it was necessary to subdivide lineages to escape an infinite regression in classification. Simpson suggested that subdivision of lineages was necessary because if it were not done, one could go from man to protist within the species *Homo sapiens* since one would never leave a single uninterruped lineage. Yet, as Hennig (1966), Bonde (1975), and Wiley (1978) have asserted, genealogical descent is composed of continua (lineages/species) and splitting of those continua. As long as we classify by nature's own history of lineage splitting and not the psychologism we impose on nature, no such infinite regression will occur. This is because (a) higher taxa arise concurrently with their own ancestral stem species (Hennig, 1966; Wiley, 1977) and (b) classifications document past continua as long as they are based on monphyletic taxa (*sensu* Hennig, 1966). Although it is true that one can run from man to protist in one taxon, that taxon would be Eucaryota, not *Homo sapiens*. There was a clade *Homo* before there was a *Homo sapiens*, just as there was a Eucaryota before any Recent protist or *Homo sapiens*.

Taking this concept of species, we might now ask, Who is inflating the number of species? The phylogeny of some Plesiadapidae Gingerich presents in this volume has 19 named "species," but only 7 lineages, 3 hypothesized ancestral and 4 hypothesized terminal. How many of these species would Eldredge and Gould recognize? I don't know. They have not discussed this particular problem, nor have they clearly formulated a species concept that would answer the question. However, Gould and Eldredge (1977, p. 117) have asserted that species are real entities in nature, and because they accept a "biological" species concept (their quotation marks) I would conclude that they would recognize 7 species. Whether one accepts a biological species concept such as Mayr's (1963) or an evolutionary species concept (*sensu* Wiley, 1978), one is forced as a logical consequence to reject the recognition of allochronic or sequential "species" (see Mayr, 1964, p. 153, and discussion by Wiley, 1978).

# Evolutionary Species and Phylogeny Reconstruction

I will now examine the implications that the evolutionary species concept has for phylogeny reconstruction. Besides reducing the number of recognized species in paleontological studies, the concept has rather far-reaching implications for understanding the relationship between cladograms and phylograms (phylogenetic trees). I was first made aware of the differences, or supposed differences, between cladograms and phylograms by Gareth Nelson (unpublished manuscript). Unfortunately, his manuscript has not yet been published, but the essence of it has been discussed by both Cracraft and Eldredge (in this volume) and by Platnick (1977b) and Tattersall and Eldredge (1977).

The basic thesis of these authors is that there are several potential or possible phylograms for each cladogram and that cladograms do not imply evolutionary events until translated into phylograms. (As Eldredge, in this volume, has delightfully put it, cladograms are not "evolutionary event-o-grams.") For example, figure 1a is a cladogram of the relationship between taxa A, B, and C. It says that A and B share two synapomorphies not shared by C (1' and 2') and that A, B, and C form a monophyletic group

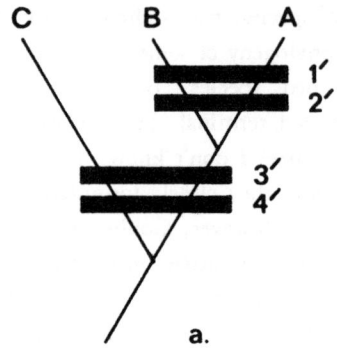

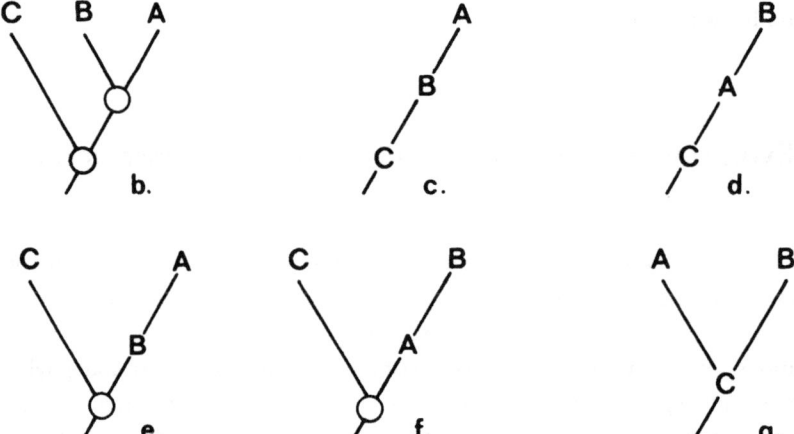

Figure 1. (a) Cladogram showing the relationship among taxa A, B, and C. (b–g) Alternative phylograms for taxa A, B, and C.

corroborated by two additional synapomorphies (3' and 4'). From this, we are supposed to be able to derive several phylograms. These are shown in figures 1b–1g. (For the purpose of simplifying the discussion, I will not consider the other three possible cladograms of the interrelationships of A, B, and C.)

Are all of these phylogenies viable alternatives at any level of the hierarchy? In other words, are all of these phylogenies possible phylogenies whether taxa A, B, and C are species, genera, or families? Tattersall and

Eldredge (1977) and Platnick (1977b) imply that they are, since they use the term taxon in an unrestricted manner. I would suggest that this is not so and that if A, B, and C are genera, for example, only one phylogeny is a viable alternative to the cladogram. This phylogeny is figure 1b, the exact reflection of the cladogram except that circles are placed at the branching points (a trivial difference in my opinion). All other "alternative" trees contain supraspecific ancestors, and since these "ancestors" are neither justifiable sets nor real biological entities, the phylogenies containing them are not viable alternative phylogenies to the cladogram. (A monophyletic supraspecific taxon *sensu* Hennig (1966) is automatically rejectable as an ancestor because its included species shares supposed descendants.) For genus B to the "ancestor" of genus A (figs. 1c and 1e), genus B would have to be a paraphyletic group, and such groups are inadmissible. The same fault holds to a greater or lesser degree with all alternatives (figs. 1c–1g) except figure 1b. I would conclude that there is only one phylogenetic tree for each cladogram of the relationships between supraspecific taxa.[4] The tree and cladogram both portray the evolutionary events connected with the origin of the taxa involved; i.e., the cladogram and phylogram are identical, and both hypothesize two speciation events.

Let us now say that A, B, and C are three evolutionary species as defined by Wiley (1978). Given this, is tree 1c a logical alternative to cladogram 1a? No, because A, B, and C are hypothesized as part of the same lineage and thus are part of the same evolutionary species and not three evolutionary species. In other words, tree 1c would not represent an alternative to cladogram 1a, but a refutation of the hypothesis that A, B, and C are all evolutionary species. Also, trees 1e and 1f are not alternatives to cladogram 1a, but refutations, where there are only two hypothesized evolutionary species. No further phylogenetic analysis would be necessary in these cases, since it is a truism that two species are sister groups, given the existence of no other species. Finally, tree 1g does hypothesize three evolutionary species, but it is not an alternative to cladogram 1a, because if species C were ancestral to species A and B, then C would share all of the synapomorphies of its descendants (Engelmann and Wiley, 1977). Yet the cladogram shows two synapomorphies shared by species A and B not shared by species C, and for tree 1g

---

[4] Eldredge (in this volume) has reached the same conclusion, i.e., that supraspecific "ancestors" are not ancestors.

to be true, these "synapomorphies" would have to be shown to be nonhomologies.[5] This can only be done by accepting an alternative cladogram, and thus tree 1g is not an alternative to cladogram 1a, but to a trichotomous relationship, such as that shown in figure 2a. The conclusion is simple. If A, B, and C are hypothesized evolutionary species, then there is only one phylogram/tree for the cladogram, and that tree is an exact duplicate of the cladogram (albeit without the circles at the nodes). This would be true, I submit, even if it could be demonstrated that B was the ancestor of A, i.e., that A was originally a small peripheral isolate of B which differentiated while the major part of B remained in stasis. This is the mode of speciation said to be dominant by Mayr (1963) and Eldredge and Gould (1972; Gould and Eldredge, 1977). But to demonstrate this an investigator would have to have both vertical and lateral sample control and would have to demonstrate that "A" did in fact exist as a population of B at one time in its history. In other words, the analysis would largely be a biogeographic problem which would require (a) the establishment of B's geographical range before the speciation of A, (b) the recognition that A once had the morphology of B, and (c) the identification of the vicariance event which split A from B. But, even if this were demonstrated, it would not substantially change the phylogram presented in figure 1b; there would still be three evolutionary species and two speciation events.

Let us now consider the cladogram in figure 2a, which shows a trichotomous relationship among species A, B, and C. In this case, there are several possible phylograms (figs. 2b–2e, as outlined by Nelson), and the hypothesis that one lineage is ancestral to the other two is a viable proposition, given that the trichotomy is real and not an unresolved dichotomous situation and that one species is, in fact, an ancestor (this second requirement is necessary to dismiss the phylogram in fig. 2b). The hypotheses is, of course, impossible to demonstrate morphologically since the investigator would have to examine the $n$th synapomorphy shared by the three taxa. But a test of the hypothesis may be accomplished by biogeographic analysis.

[5] It is possible that C is a sample of the evolutionary stem species of A and B at a point where the autapomorphies it will leave to its descendants (1' and 2') have either not arisen or are so rare that they are unsampled (i.e., the frequency of individuals with 1' and 2' is low). If so, then one would expect further sampling at a later time horizon to yield a polymorphic sample. Then one would be faced with the task of corroborating the hypothesis that the derived members of the population are the same species as the primitive members of the population, rather than the alternative hypothesis that the derived individuals represent a different species that is displacing the more primitive species via interspecific competition.

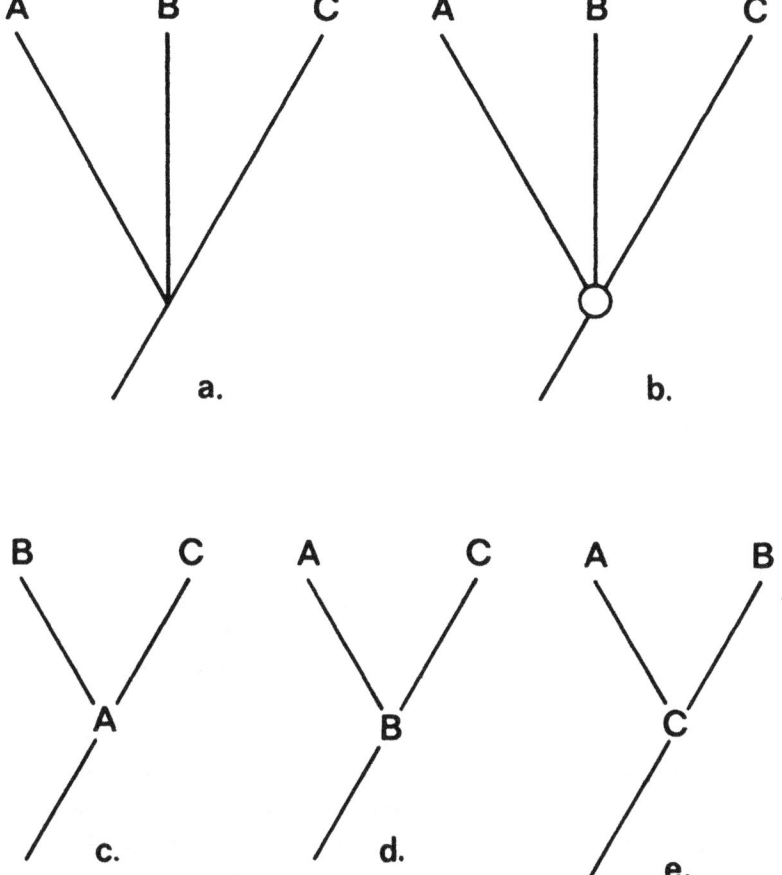

Figure 2. (a) Cladogram showing a trichotomous relationship among species A, B, and C. (b–e) Alternative phylograms for species A, B, and C.

Let us assume that A, B, and C are part of a biota. Further, let us say that A is a widespread species at time $t_1$ and that B and C occupy portions of the range of A at time $t_{1+x}$. Furthermore, let us assume that there has been no dispersal and that a general track analysis (via the method of Croizat, 1962) has indicated a vicariance event which more or less corresponds to the boundary between B and C at time $t_{1+x}$. We might then prefer tree 2c over the cladogram and the alternate trees, as a matter of parsimony. We might ask, however, how many times such analyses are possible in paleontology. These analyses certainly contain a number of assumptions that make them suspect, not the least of which is the acceptance of the trichotomy, and they demand lateral (geographic) as well as vertical (stratigraphic) control of the sample. Also note that neither morphology nor stratigraphy can corroborate the hypothesis directly—they are simply initial prerequisites for the hypothesis to be tested (rather like the initial prerequisite that A, B, and C must form a monophyletic group before their interrelationships can be critically assessed). This same situation might be extended to a single species in a trichotomous situation with two branches, each containing several species (i.e., a monotypic genus and two polytypic genera). In this case, the same reasoning applies, although it is obviously a more complicated biogeographic situation.

Now, consider A, B, and C in figure 1a as unspecified morphotypes which may be species or populations of a single species.[6] In other words, the investigator does not wish to form a hypothesis concerning the number of lineages in the samples. In this case, "trees" 1b through 1f are viable alternatives to cladogram 1a. This may be an attractive approach to many paleontological problems. The analysis proceeds from a cladogram of unspecified morphotypes to a cladogram/tree of a specified number of evolutionary species/lineages. But once that determination is made, the cladogram is a tree and an "evolutionary event-o-gram." And the number of evolutionary species at the end of the analysis may or may not equal the number of morphotypes with which the investigator began. If the investigator hypothesized tree 1c, with a single evolutionary species composed of three samples, then two predictions can be made: that additional sampling will reveal several populations composed of polymorphisms for synapomorphies 1' and 2' and their plesiomorphic counterparts between sample populations C and B and that these synapomorphies will be fixed from sample B to sample A. Of course, both predictions imply gradualistic fixation of synapomorphies within the lineage, an interesting test case for phyletic gradualism (i.e., do synapomorphies arise via phyletic gradualism, saltation, or "quick" selection concurrent with speciation events?).

I draw the following conclusions from the above discussion. (1) Supraspecific taxa cannot be ancestors at all and therefore supraspecific ancestors do not belong in any natural system of phylogeny reconstruction or classification. (2) Lineages are the largest natural groupings in evolution that

---

[6] Supraspecific groups cannot be unspecified morphotypes by the very nature of their demonstrated or presumed monophyly.

are subject to the same evolutionary or homeostatic forces as a unit, and species are best thought of as lineages. (3) In phylogenetic reconstructions of supraspecific taxa, the number of possible phylogenetic trees is equal to the number of possible cladograms for any given set of supraspecific taxa. (4) In phylogenetic reconstructions involving evolutionary species, the number of possible phylogenetic trees is equal to the number of possible cladograms for any particular group of species, given that the number of evolutionary species is correctly assessed and given a dichotomous cladogram. (5) In the case of trichotomous or multichotomous cladistic geometries of species or mixed species/supraspecific taxa, the number of phylogenetic trees is not equal to the number of cladograms, given a true nondichotomous phylogram of the taxa (i.e., that for a three-species or one-species/two-genera situation, only one speciation event has occurred in nature at the level of universality considered).

I suggest that the evolutionary species concept is best suited for elucidating nature's own history of speciation because that history is one of the splitting of lineages (and indeed, this may be where much of the evolutionary "action" is; see Gould and Eldredge, 1977). Because it is the best concept for reconstructing the pattern of evolution, it also provides the best basis for deducing the *mode* of evolution. Indeed, how can we profitably study the process of evolution without a notion as to the pattern it has produced?

# References

Ashlock, P. H. 1971. Monophyly and associated terms. *Syst. Zool.* 20:63–69.
Ball, I. R. 1975. Nature and formulation of biogeographic hypotheses. *Syst. Zool.* 24:407–30.
Bonde, N. 1975. Origin of "higher groups"; Viewpoints of phylogenetic systematics. *Problèmes actuels de paléontologie—évolution des vertébrés; Colloq. Int. CNRS* 218:293–324.
Brundin, L. 1966. Transantartic relationships and their significance, as evidenced by chironomid midges. *Kungl. Svenska Vetenskapakad. Handl.* 11:1–472.
Brundin, L. 1968. Application of phylogenetic principles in systematics and evolution. In: T. Orvig (ed.), *Nobel Symposium 4: Current Problems in Lower Vertebrate Phylogeny,* pp. 473–95. Almqvist and Wiksell, Stockholm.

Colless, D. H. 1972. A note on Ashlock's definition of "monophyly." *Syst. Zool.* 21:126–28.
Cracraft, J. 1974. Phylogenetic models and classification. *Syst. Zool.* 23:71–90.
Croizat, L. 1962. Space, time, and form: The biological synthesis. Published by the author, Caracas.
Eldredge, N., and Gould, S. J. 1972. Punctuated equilibria: An alternative to phyletic gradualism. In: T. J. M. Schopf (ed.), *Models in Paleobiology,* pp. 82–115. Freeman, Cooper, and Co., San Francisco, Calif.
Engelmann, G. F., and Wiley, E. O. 1977. The place of ancestor–descendant relationships in phylogeny reconstruction. *Syst. Zool.* 26:1–11.
Farris, J. S. 1974. Formal definitions of paraphyly and polyphyly. *Syst. Zool.* 23:548–54.
Ghiselin, M. T. 1966. On psychologism in the logic of taxonomic controversies. *Syst. Zool.* 15:207–15.
Ghiselin, M. T. 1974. A radical solution to the species problem. *Syst. Zool.* 23:536–44.
Gould, S. J., and Eldredge, N. 1977. Punctuated equilibria: The tempo and mode of evolution reconsidered. *Paleobiology* 3:115–51.
Hennig, W. 1966. *Phylogenetic Systematics.* University of Illinois Press, Urbana.
Hennig, W. 1975. "Cladistic analysis or cladistic classification?": A reply to Ernst Mayr. *Syst. Zool.* 24:244–56.
Hull, D. L. 1976. Are species really individuals? *Syst. Zool.* 25:174–91.
Lewontin, R. C. 1974. *The Genetic Basis of Evolutionary Change.* Columbia University Press, New York.
McKenna, M. C., Engelmann, G. F., and Barghoorn, S. F. 1977. "Review of Cranial Anatomy and Evolution of Early Teritary Plesiadapidae (Mammalia, Primates)" by P. D. Gingerich. *Syst. Zool.* 26:233–38.
Mayr, E. 1963. *Animal Species and Evolution.* Belknap/Harvard Press, Cambridge, Mass.
Mayr, E. 1964. *Systematics and the Origin of Species.* Dover Publications, New York.
Nelson, G. J. 1971. "Cladism" as a philosophy of classification. *Syst. Zool.* 20:373–76.
Platnick, N. I. 1977a. Paraphyletic and polyphyletic groups. *Syst. Zool.* 26:195–200.
Platnick, N. I. 1977b. Cladograms, phylogenetic trees, and hypothesis testing. *Syst. Zool.* 26:438–42.
Raup, D. M., Gould, S. J., and Schopf, T. J. M. 1973. Stochastic models of phylogeny and the evolution of diversity. *J. Geol.* 81:525–42.
Simpson, G. G. 1961. *Principles of Animal Toxonomy.* Columbia University Press, New York.
Tattersall, I., and Eldredge, N. 1977. Fact, theory, and fantasy in human evolution. *Amer. Sci.* 65:204–11.
Wiley, E. O. 1976. The phylogeny and biogeography of fossil and recent gars (Actinopterygii: Lepisosteidae). *Misc. Publ. Univ. Kansas Mus. Nat. Hist.* 64:1–111.

Wiley, E. O. 1977. Are monotypic genera paraphyletic? A reply to Norman Platnick, *Syst. Zool.* 26:352–55.
Wiley, E. O. 1978. The evolutionary species concept reconsidered. *Syst. Zool.* 27:17–26.

# Index

Acastacea, 173, 176–77, 180, 192
*Acastava*, 180
*Acastella*, 180
*Acastocephala*, 180
*Acastoides*, 177, 179–80
Adanson, M., 12
Adapidae, 55
*Adapis*, 56
Adaptation, 24, 28, 168, 192–93, 209
Adaptive zone, 27
*Ad hoc* hypothesis, 98–100
Advanced character, see Derived characters
Ager, D. V., 187
Allen, J. A., 136
Allopatric speciation, 8, 23–25, 36, 113, 138–39, 143, 145, 184; see also Speciation
Allopatry, 23, 204
Amniota, 94
Amphibian–reptile transition, 66
Anagenesis, 62–63
Ancestor–descendant relationships (hypotheses), 29, 34–35, 56, 87–89, 113, 115, 130, 134, 137, 141–42, 144, 148, 150–56, 166, 183, 188, 201, 214, 216; biogeographic criterion, 220–22; among supraspecific taxa, 118, 222
Ancestor–descendant species, 3, 32, 56, 59, 87, 113–14, 118, 149, 168, 182, 190, 211
Ancestral character-states, 149; see also Character states
*Anchomomys*, 55–56
*Archaeopteryx*, 66, 69, 74
Aristotelianism, 10
Ashlock, P., 212
*Astarte*, 149
*Australopithecus*, 72–73

Autapomorphy, 34, 86, 186
Auxiliary impression system of trilobites, 174, 176
Axioms of phylogenetic system, 85

Baconian view of science, 80
Baird, D., 64
Baldis, B., 194
Ball, I. R., 213
Bambach, R. K., 119, 127–28
Barghoorn, S. F., 216
Basic taxa, 89, 92
*Bellucina*, 152
Biogeography, 7, 203, 205; analysis, 205, 208; unit, 207
Biological species concept, see Species
Biostratigraphy, 50, 55, 131
Blackwelder, B. W., 121
Bock, W. J., 30, 84, 113–14
Body size, 44–45, 73
Boehner, P., 98
Bonde, N., 80, 85, 89, 103, 150, 213–14, 216
Boucot, A. J., 4, 151, 204
*Bouleia*, 186
Bourque, P. A., 185
Boyd, D. W., 117
Branisa, L., 178, 186
Bretsky, S., 3, 30, 116–17, 125–26, 132, 135, 141, 148, 152, 154, 172, 202, 204–5, 207, 212
Briggs, J. C., 138
Britton, J. C., 121, 152–54
Brongniart, A., 42
Brundin, L., 86, 166, 213
Buddenhagen, C. H., 149

Buffon, G. L. L., comte de, 12, 17
Burkhardt, R. W., Jr., 12
Bush, G., 27, 184
Butterfield, H., 10

*Calmonia*, 179–80
Cameron, H. D., 92
Campbell, K. S. W., 147, 166, 171, 181, 185
*Canis*, 43–45, 50
Carroll, R. L., 64
Carson, H. L., 137–38, 144
Catastrophism, 10
*Cercamonius*, 55
Change, gradual, *see* Gradualism; Phyletic evolution
Character, 57, 92, 95–96, 99–100; analysis, 3, 31, 46, 57–58, 101; contradiction, 96, 99; correlation, 148; displacement, 24, 63; distributions, 96; phylogeny, 101–2, 145; reversal, 87–88, 95, 185; states, 57, 101, 145, 149; weighting, 204
Chronocline, 46, 116
Cifelli, R., 126
Clade, 80
Cladism, 3, 80, 165–66, 185; *see also* Phylogenetic systematics
Cladistics, 143–44, 147, 199, 201–3, 205, 209; analysis, 30–32, 170, 209, 217–23; classification, 103; methods, 84; *see also* Phylogenetic systematics
Cladists, 116, 166, 200, 203–4; *see also* Phylogenetic systematists
Cladogenesis, 62–63, 80, 204
Cladogram, 1, 30–34, 89, 103, 130, 147, 155, 166–69, 172–73, 182, 184, 190, 195, 200, 212, 217, 219; analysis, 170
Clark, W. E. LeGros, 167
Clarkson, E. N. K., 174
Classification, 7, 12, 103, 113, 199, 200
Cline, and clinal variation, 46, 57, 145
Choanata, 92
Cobban, W. A., 149
*Codakia*, 134–36, 148
Coelurosaur, 66
Colbert, E. H., 116
Coleman, W., 13, 18
Colless, D. H., 213
Colonization, 135

Commonality principle, 170
Common ancestry, 29, 87–88, 96; *see also* Ancestor–descendant relationships
Community, 203, 205, 208; ecology, 209; group, 207–8; group analysis, 209
Competition, interspecific, 220
Conrad, T. A., 152
Convergence, 95, 99, 167, 201
Cook, H. E., 128
Cope, E. D., 59
*Copelemur*, 62
Coppens, Y., 70
Corroboration, of phylogenetic hypotheses, 83; *see also* Hypothetico-deductive method
Cosmopolitan, 208
Cracraft, J., 3, 29, 32, 99, 113–17, 130, 137, 139, 141, 143, 147, 150, 202, 206, 213, 217
Croizat, L., 139, 221
Crompton, A., 68
Crossopterygian, 63
Crovello, T., 43
Crowson, R., 103, 114, 131
*Cryphaeoides*, 180–81
Cuvier, G., 11, 13, 42, 58

Dall, W. H., 152
Dampier, W. C., 74
Darwin, C., 2, 8–9, 13–14, 16–17, 23–24, 29, 42, 58
Day, M. H., 70
Deductive method, 81; *see also* Hypothetico-deductive method
Dennison, J. M., 135
Derived characters, 3, 31, 85–86, 170, 200–201; distributions of, 84
Derived similarities, 31
Descendant, 59, 149; *see also* Ancestor–descendant relationships
Diamond, J., 154
Diapsida, 94
Distributional data in systematics, 172
Diversity, 121, 125, 127; through time, 126–29
Dobzhansky, T., 19
Dodson, P., 149
Dogs, 43; see also *Canis*
Drosophilidae, 144

# Index

DuBar, J. R., 121
Durham, J. W., 117, 119, 126, 134, 137

Edwards, J. L., 99–101, 173
Einstein, A., 81
Eldredge, N., 3, 9, 18, 20–25, 28–30, 33, 58–59, 62, 87–88, 101, 103, 114, 116–17, 119, 130, 137–39, 141–45, 147, 152, 155, 166, 168–70, 172, 174, 177–78, 182, 184–86, 190, 200, 202–4, 206, 212, 214–15, 217, 219, 220, 223
Elzanowski, A., 66
Endemicity, 208
Engelmann, G. F., 87, 118, 172, 186, 188, 213, 216, 219
Eocene, 46, 55, 115
Equilibrium of species through time, 61
Essentialism, 11–12, 17
Evolution, 8, 14, 17, 85, 147, 204, 209; mechanism, 87–88, 100; models, 7; modes, 205–6, 215, 223; natural selection, 100; paleontology, 35; process, 4, 8–9, 22, 166, 216; quantum, 27, 205–6, 209; theory, 3–4, 42, 86–87, 144; zigzag, 204; *see also* Phyletic evolution; Scenario, evolutionary; Species
Evolutionary systematics, 4, 30, 114–16, 141, 144–45, 151, 154
Evolutionary trees, *see* Trees, evolutionary
Explanation, 82
Extrinsic data, 3

Falsification, 34, 81, 83, 92, 98, 186; *see also* Hypothetico-deductive method
Farris, J. S., 131, 145, 151, 170, 213, 217
Faunal succession, principle of, 50
Fischer, A. G., 126
Flessa, K. W., 151
Fossilization, 119
Fossil record: Darwin's view of, 16; gaps in, 119, 129, 200
Frazetta, T. H., 147
Functional complexes, role of in systematics, 100

Gaffney, E., 3, 30, 33, 92, 94, 181, 199, 214
Gale, B. G., 11, 18
*Galesaurus*, 69

Gandl, J., 175
Genealogical descent, 216
Genealogy, 87-
Generality, levels of, 85; *see also* Universality, levels of
Genetic homeostasis, 59
Genetic similarity, 114
Geographic distributions, 156, 172, 185, 188
Geologic time, 202
George, R. Y., 128
Ghiselin, M., 23, 89, 214
Gingerich, P. D., 3, 22, 27, 30, 59, 66, 69–70, 126, 166, 172, 188, 190, 202–3, 205, 207, 215, 217
*Gobipteryx*, 66
Goldschmidt, R., 58
Goodfield, J., 12–13
Gould, S. J., 9, 18, 20–25, 28, 45, 58–59, 62, 114, 117, 130, 137–39, 141–42, 147–48, 151, 154, 190, 204, 214–15, 217, 220, 223
Gradations, 12
Gradualism (gradualistic philosophy), 11, 13, 17–19, 24–26, 28–29, 48, 206; *see also* Phyletic evolution
Greene, J. C., 10
Griffiths, G. C. D., 151

Haeckel, E., 2
Hallam, A., 138
Hansen, T., 131
Harper, C. W., Jr., 34, 129, 153, 204
Hay, W. W., 135
Hazel, J. E., 130–32
Hecht, M. K., 22, 99–101, 103, 113, 116, 119, 166, 168, 170, 172–73
Hedgpeth, J., 209
Hempel, C., 98
Hennig, W., 3–4, 31, 36, 79–80, 84, 86, 89, 101, 103, 114, 117, 139, 142, 149, 166–67, 169, 199–200, 213, 216, 219; Hennigism, 166; *see also* Cladistics
Hominidae, 69–73; *Australopithecus*, 72–73; *Homo erectus*, 72; *Homo sapiens*, 69, 72; phylogeny (evolution), 69, 70, 72; *Zinjanthropus*, 70, 72
Homology, 31, 95–96, 170, 181; determination of, 96
Hopson, J., 68

Hull, D. L., 11, 23, 80, 84, 214
Hume, D., 81
Hutchinson, G. E., 45
Hutton, J., 42
Huxley, J., 19, 46
Hypothesis, 81, 85, 92, 172, 182, 194; creation of, 81; least-rejected, 94, 96, 99 (see also Parsimony); testing, 33, 55, 81–83, 92, 170
Hypothetico-deductive method, 29, 33, 36, 80–81, 84–85, 98; see also Falsification

*Ichthyornis*, 66
*Ichthyostega*, 63–64, 69, 74
Imbrie, J., 151
Individual (philosophical), 89
Induction, 81
Infinite regress, 83, 216
Inheritance, 87
Intrinsic data, 3

Jackson, J. B. C., 131, 135–36
Jeletzky, J. A., 134
Jenkins, F. A., 68

Kanamori, H., 138
Kauffman, E., 125, 130, 132, 135, 149
Keen, A. M., 138
Kennedy, W. J., 149
Kier, P. M., 119, 121, 126
*Kloucekia*, 175
Kluge, A., 170
Kohlberger, W., 98
Kuhn, T., 80, 117

Lamarck, J. B., 13, 42, 58
*Latimeria*, 201
Leakey, R., 70, 72–73
Lespérance, P. J., 185
Levinton, J. S., 116
Lewontin, R. C., 130, 213
Lineages, 46, 62, 89, 214
Links, evolutionary, 14; see also Transitional forms
Linnaeus, 12
Lithostratigraphy, 50, 55
Lovejoy, A., 9, 11–12, 14, 17

Lucinidae, 121, 126–28, 132, 152; *Lucina*, 152, 154
Lyell, C., 13, 18, 42

MacArthur, R. H., 45
McKenna, M. C., 103, 138, 188, 216
McNab, B. K., 45–46
macroevolution, 24, 58, 147
Magee, B., 81
Mammal-like reptiles, 69
Mammals, 41
Martin, L. D., 66
Maslin, T. P., 117
Matthew, W. D., 59–60
Mayr, E., 11, 19–20, 26, 30, 36, 41, 43, 48, 79–80, 87, 103, 114, 116, 138, 141, 167, 217, 220
Menzies, R. J., 128
*Metacryphaeus*, 179–80, 186, 193
Microevolution, 147
Microgenesis, 58
Miles, R., 84
Modes, evolutionary, 205–6, 215, 223
Mollusks, Tertiary, 113–163 *passim*
monophyletic groups, 80, 213, 217
Monophyly, 87–89, 92, 150, 182, 212–13
Moore, R. C., 130
Moores, E. M., 139
Morgan, E., 155
morphocline, 58, 101–2, 142, 149, 152; polarity, 117, 149, 155 (see also Polarity)
morphological: discontinuity, 142; postulate; stasis, 26; transformation, 137 (see also Morphocline polarity)
morphotype, 145

natural selection, 100
Nelson, G. J., 30, 36, 80, 85, 94, 103, 113, 131, 139, 166, 169, 183, 213, 217, 220
*Neopilina*, 201
Newell, N., 117
New Systematics, 7, 19, 23
nonhomology, 31; see also Homology
*Notharctus*, 56, 60
numerical taxonomy, 2, 4, 30, 114

Oaks, R. Q., Jr., 121
Ockham's razor, 97–98; see also Parsimony

Olsson, A. A., 153
Ontogenetic: changes, 148; events, 148; sequences, 102
Ontogeny, 152, 154, 156, 171, 180, 203
organic diversity, 85
origin: of birds, 66; of mammals, 68; of species, 14, 22, 113–14; *see also* Species; Speciation
*Ormathops*, 175
Ormiston, A., 177–78
Osborn, H. F., and hypothesis of gradual appearance, 19
Ostrom, J., 66
Outgroup comparison, 92, 94, 101–2, 172
Overall similarity of taxa, 114

Paleontology, 7, 9, 22–23, 25, 165
Palmer, A. R., 201
*Parabouleia*, 179, 186
*Paradoxides*, 201
Parallelism, 94–96, 99–100, 167, 170, 172–73, 201, 204; *see also* Convergence
Paraphyly and paraphyletic groups, 86, 211, 213
Parrington, F. R., 68
Parsimony, 34, 96–99, 172–73, 181, 185
*Parvilucina*, 152
Patterson, C., 95–96
*Pelycodus*, 46, 49–50, 55–56, 59–63, 70, 73
*Pennaia*, 180
*Periconodon*, 56
Peripheral isolate, 138, 143; *see also* Speciation
*Phacoides*, 133
Phacopid trilobites, 173, 187, 200
*Phacopina*, 174, 177, 179, 193
*Phacops*, 180, 185
Phenetics, 114, 116, 145, 155 (*see also* Numerical taxonomy); clusters (species), 43; linking, 54; similarity, 57, 114
Philosophers, 10, 201, 209; *see also* Einstein, A.; Hull, D. L.; Kuhn, T.; Popper, K.
Phyletic evolution (phyletic gradualism), 17–18, 20–22, 24, 26, 46, 48, 58–59, 61–62, 73, 113–15, 131, 134, 137, 141, 144, 150, 166, 184, 190, 205, 215, 222
Phylogenetic, 84, 192, 212; analysis (phylogenetic reconstruction), 3–4, 7–9, 33, 41, 49, 55, 63, 84, 87–88, 99, 113, 115, 119, 147, 165–66, 217; axioms, 85; evidence, 115; history, 117; hypotheses, 85–86; models, 4, 206; relationships, 1, 117, 166; systematics, 3, 8, 30, 35–36, 80, 116–17, 142–43, 145, 147, 150, 155, 166, 199–200 (*see also* Cladism; Cladistics; Hennig); systematists, 32, 113–15, 130–31, 137, 142–43, 145, 149 (*see also* Cladists); trees, 1, 34, 89, 168, 172, 182, 190, 193, 195, 217, 223 (*see also* Trees, evolutionary)
Phylogeny, 3, 8, 41–42, 54–55, 87, 147, 171, 199, 204, 209, 211; empiricism, 73; higher-level, 94
Phylogram, 217; *see also* Cladogram
Platnick, N. I., 34–35, 86–87, 92, 99, 101, 139, 141–42, 170, 181, 186, 213
Plesiomorphy, 86, 213; *see also* Primitive characters
*Pleurolucina*, 152, 154
Polarity, 58, 102, 148; *see also* Morphocline, polarity
Polyphyletic, 86
Popper, K., 79, 81–82, 84, 96, 98–99
Prediction, 82
Primitive characters, 3, 31, 170, 200–201; *see also* Plesiomorphy; Symplesiomorphy
*Probolops*, 181
*Pronothodectes*, 56
*Pronycticebus*, 55
*Protoadapis*, 55
Punctuated equilibrium, 25–28, 58–59, 61, 62, 73, 114, 134, 137, 141, 143–44, 190, 215; *see also* Speciation

Quantum evolution, 27, 205–6, 209

*Radiolucina*, 152
Ratite birds, 206
Raup, D., 119, 214
Refutation: of hypotheses, 83; potential, 170; *see also* Hypothesis testing
Reptiles, 64, 68, 69
Richter, E., 180
Roe, A., 130
Romer, A. S., 64
*Romeriscus*, 64, 69, 74
Rosen, D. E., 138–39
Ross, H. H., 26, 154
Roughgarden, J., 137, 139

Rowe, A., 128
Ruse, M., 23

Saltation, 17, 19, 62, 184, 222
Sauropsida, 94
Säve-Söderbergh, G., 63
Scenario, evolutionary, 155, 168–69, 172, 192–93, 195
Schaeffer, B., 63–64, 101, 103, 113, 116, 119, 166, 168, 170, 172
Scheltema, R. S., 131
Schindewolf, O. H., 58
*Schizostylus*, 179
Schoeninger, M. J., 69–70
Schopf, T. J. M., 128, 214
Selection, 27, 100, 193, 222
*Seymouria*, 64
Shared derived characters, *see* Synapomorphy
Shared primitive characters, *see* Symplesiomorphy
Shergold, J. H., 175, 180
Simons, E. L., 59
Simpson, G. G., 19–20, 23, 25, 27, 30, 35–36, 41, 46, 58, 103, 114, 116–17, 130, 138, 144, 147, 150, 155, 167, 205–6, 211–12, 216
Singular scientific statements, 96
Sister group of taxa, 171–72, 213
Sloan, R., 192
Smith, W., 11, 42
Sneath, P. H. A., 30, 43, 48, 57–58, 79, 114
Sokal, R., 30, 43, 48, 57–58, 79, 114
Special creation, 86
Speciation, 7, 20, 26–27, 58, 86, 88–89, 137, 141–43, 145, 150, 154, 166, 184, 220, 222–23; *see also* Allopatric speciation; Punctuated equilibrium
Species, 23, 29, 43, 48, 88, 182, 217, 223; arbitrary delineation, 48, 61; concepts, 215; discrete, 15–17, 23, 26; diversity, 118–19, 121, 126; duration, 136; equilibrium, 61; evolutionary, 211, 217–23; fixity, 12, 144; ranges, 118, 129, 131; reality, 8
Sprinkle, J., 151
Stanley, S., 45, 135–36
Stegocephalia (amphibians), 63
Stensiö, E., 187
Stenzel, H. B., 115

Stewart, J. D., 66
Stratigraphic, 156, 172, 202; distribution (range), 130, 132, 156, 185, 188; position, 115, 117, 130, 186–87, 201; relationship, 148, 152; sequence, 117, 127, 137, 155
Stratophenetic approach, 41, 49, 54, 56, 63, 69–70, 73, 127, 166, 188, 207
Struve, W., 175, 177
Sullivan, W., 138
Supraspecific taxa (groups), 150–51, 153–54, 214–15, 222–23; ancestors, 211–12, 219, 222 (*see also* Ancestors)
Sylvester-Bradley, P. C., 21, 141
Symplesiomorphy, 31, 114, 172, 181
Synapomorphy, 31, 85–86, 92, 95–96, 102, 113, 155, 167, 169–70, 172–73, 176, 181–82, 185–86, 200, 213, 220
Synthetic theory of evolution, 86–87
Szalay, F., 34

Takeuchi, H., 138
*Tarijactinoides*, 181
Tattersall, I., 87, 114, 145, 152, 155, 166, 169, 184, 217–18
Tautology, 82–83
Taxa, 85–86; *see also* Species; Supraspecific taxa
Taxonomy, 199–202, 204–5, 209; *see also* Numerical taxonomy
Taylor, J. D., 136
Taylor, M. E., 128
Testing, *see* Hypothesis testing
Tetrapoda, 94
Therapsida (reptiles), 68
Thompson, I., 126–28
Thomson, K. S., 64
Three-taxon statement, 89, 98
Toulmin, S., 12–13
Transitional forms: *Archaeopteryx*, 66–68; *Ichthyostega*, 63–64; *Romeriscus*, 64, 66
Trees, evolutionary, 31, 33, 88, 166, 168–69, 172, 182, 200–203, 206, 208–9; A-trees, 183; construction of, 209; X-trees, 33–34, 183, 189; *see also* Phylogenetic trees
Tree analysis, 32, 188–89
Trichotomy, in testing hypothesis, 94
Trilobita, 165–98, *passim*
Trueman, A. E., 144
Typology, 61, 215

Uniformitarianism, 10
Universality, levels of, 82, 92, 96, 100
Uyeda, S., 138

Valentine, J., 119, 128, 139, 147, 151
Van Valen, L., 45, 192
Variation, 207
Vertebrata, 94
Vertebrate Phylogeny, 41
Vicariance, 139, 142, 145, 150
Vorzimmer, P. J., 16

Walker, A. C., 70, 72
Wellnhofer, P., 66
White, T. D., 70
Wiley, E. O., 4, 33–34, 87, 94, 96, 118, 130, 186, 211, 213–14, 216–17, 219
Wolfe, N., 82

*Zinjanthropus,* 70, 72
Zittel, K. A. von, 18

Bei Fragen zur Produktsicherheit wenden Sie sich bitte an:
If you have any questions regarding product safety,
please contact:

Walter de Gruyter GmbH
Genthiner Straße 13
10785 Berlin
productsafety@degruyterbrill.com